Thin Film Device Applications

Thin Film Device Applications

KASTURI LAL CHOPRA
AND
INDERJEET KAUR

Indian Institute of Technology
New Delhi, India

PLENUM PRESS • *NEW YORK AND LONDON*

Library of Congress Cataloging in Publication Data

Chopra, Kasturi, L., 1933–
  Thin film device applications.

  Bibliography: p.
  Includes index.
  1. Thin film devices. I. Inderjeet Kaur. II. Title.
TK7872.T55C48 1983   621.381'71   83-9632
ISBN 0-306-41297-7

©1983 Plenum Press, New York
A Division of Plenum Publishing Corporation
233 Spring Street, New York, N.Y. 10013

Printed in the United States of America

Dedicated
to the Cause of
Science and Technology
in the Service of
Mankind

Preface

Two-dimensional materials created *ab initio* by the process of condensation of atoms, molecules, or ions, called thin films, have unique properties significantly different from the corresponding bulk materials as a result of their physical dimensions, geometry, nonequilibrium microstructure, and metallurgy. Further, these characteristic features of thin films can be drastically modified and tailored to obtain the desired and required physical characteristics. These features form the basis of development of a host of extraordinary active and passive thin film device applications in the last two decades. On the one extreme, these applications are in the submicron dimensions in such areas as very large scale integration (VLSI), Josephson junction quantum interference devices, magnetic bubbles, and integrated optics. On the other extreme, large-area thin films are being used as selective coatings for solar thermal conversion, solar cells for photovoltaic conversion, and protection and passivating layers. Indeed, one would be hard-pressed to find many sophisticated modern optical and electronic devices which do not use thin films in one way or the other.

With the impetus provided by industrial applications, the science and technology of thin films have undergone revolutionary development and even today continue to be recognized globally as frontier areas of R/D work. Major technical developments in any field of science and technology are invariably accompanied by an explosion of published literature in the form of scientific publications, reviews, and books. So vast is this rapidly expanding field that it is not humanly possible to write a comprehensive treatise on the subject in a single volume, and thus only specialized monographs and edited reviews appear in the literature. *Thin Film Phenomena*, by K. L. Chopra (1969), and *Handbook of Thin Film Technology*, edited by L. Maissel and G. Glang (1970), provide good early reviews of the science and technology of thin films, along with a limited description of thin film device applications. Specialized applications of thin films are covered in such books as *Science and Technology of Surface Coatings*, edited by B. N. Chapman and J. C. Anderson (1973); *Thin Film Optical Filters*, by H. A. Macleod (1969); *Active and Passive Thin Film Devices*, edited by J. C. Coutts (1978); *Polycrystalline and Amorphous Thin Films and Devices*, edited by L. Kazmerski (1980); and *Thin Film Solar Cells*, by K. L. Chopra and S. R. Das (1983). The need for a single, concise, cohesive, and comprehensive textbook on thin film device applications clearly exists and has been felt by graduate students and workers in the field. We were inspired by Prof. D. S. Campbell to fill this gap, and the result is this monograph. Though overly concise, this book provides a self-contained coverage of a wide range of device applications of thin films in such areas as optics, electro-optics, microelectronics, magnetics, quantum engineering, surface engineering, and thermal detection.

What thin films are and how these are prepared and characterized forms the Introduction (Chapter 1). The role of thin films in a wide range of optical and electro-optical applications is described in Chapters 2 and 3. This is followed by the micro-electronic and magnetic applications in Chapters 4 and 5. The superconducting properties and quantum tunneling effects of thin films are the basis of a host of quantum engineering applications which are the subject of Chapter 6. Thermal applications of thin films are discussed in Chapter 7. Finally, a whole range of passivating, tribological, decoration, and biomedical applications form the subject matter of Chapter 8.

The emphasis in all chapters is to give a brief physical basis and technical description of the devices in a cohesive manner. Adequate references are provided for the interested reader to dig deeper into the details of the devices. Both academically interesting and commercially viable devices have been covered to inspire the innovative mind. Besides serving as a textbook for graduate students in applied sciences and engineering, it should be a good reference book for scientists and engineers involved in R/D work on thin-film-based devices.

A book of this type draws heavily and unhesitatingly from the published works of numerous authors. We have made a concerted effort to describe representative examples of applications of thin films from the literature in various areas in a very condensed and brief form. If we have made some glaring errors or omissions, we would be grateful to the reader for pointing these out to us. It is our earnest hope that this unique book will find wide acceptance as a textbook which will inspire our readers to explore the vast and virgin field of the microscience and technology of low-dimensional thin film materials.

We acknowledge the assistance of our colleagues in the Thin Film Laboratory of our Institute in various forms in the preparation of this book. We are indebted to Mr. Irfan Habib for his critical reading of the entire manuscript.

Finally, we are indebted to our families whose exemplary patience and moral support have sustained us to the completion of this work.

New Delhi   *K. L. Chopra and I. Kaur*

Contents

Chapter 3. Optoelectronic Applications *89*

Chapter 4. Microelectronic Applications *129*

Chapter 8. Surface Engineering Applications *255*

1

Thin Film Technology: An Introduction

In order to appreciate thin film device applications, it is essential to understand what thin films are, what makes them so attractive for applications, and how they are prepared and characterized. A brief review of the salient and relevant features of the topics is presented in the following sections of this chapter. For more details, the reader is referred to a host of reviews[1-6] and books[7-12] on the subject written from different viewpoints.

1.1. Why Thin Films?

A solid material is said to be in thin film form when it is built up, as a thin layer on a solid support, called substrate, *ab initio* by controlled condensation of the individual atomic, molecular, or ionic species, either directly by a physical process, or via a chemical and/or electrochemical reaction. Since individual atomic, molecular, or ionic species of matter may exist either in the vapor or in the liquid phase, the techniques of thin film deposition can be broadly classified under two main categories: (1) Vapor-phase deposition and (2) liquid-phase/solution deposition. It should be emphasized here that it is not simply the small thickness which endows thin films with special and distinctive properties, but rather the microstructure resulting from the unique way of their coming into being by progressive addition of the basic building blocks one by one, which is more important. Films prepared by direct application of a dispersion or a paste of the material on a substrate, and letting it dry, are called, irrespective of their thickness, *thick films* and have properties characteristically different from those of *thin films*.

In thin films, deviations from the properties of the corresponding bulk materials arise because of their small thickness, large surface-to-volume ratio, and unique physical structure which is a direct consequence of the growth process. Some of the phenomena arising as a natural consequence of small thickness are optical interference, electronic tunneling through an insulating layer, high resistivity and low temperature coefficient of resistance, increase in critical magnetic field and critical temperature of a superconductor, the Josephson effect, and planar magnetization. The high surface-to-volume ratio of thin films due to their small thickness and microstructure can influence a number of phenomena such as gas adsorption, diffusion, and catalytic activity. Similarly, enhancement of superconducting transition temperature, corrosion resistance, hardness, thermopower, and optical absorption arise in thin films of certain materials having metastable disordered structures.

1.2. *Thin Film Growth Process*

Any thin film deposition process involves three main steps: (1) production of the appropriate atomic, molecular, or ionic species, (2) their transport to the substrate through a medium, and (3) condensation on the substrate, either directly or via a chemical and/or electrochemical reaction, to form a solid deposit. Formation of a thin film takes place via nucleation and growth processes. The general picture of the step-by-step growth process emerging out of the various experimental and theoretical studies can be presented as follows:

1. The unit species, on impinging the substrate, lose their velocity component normal to the substrate (provided the incident energy is not too high) and are physically adsorbed on the substrate surface.
2. The adsorbed species are not in thermal equilibrium with the substrate initially and move over the substrate surface. In this process they interact among themselves, forming bigger clusters.
3. The clusters or the nuclei, as they are called, are thermodynamically unstable and tend to desorb in a time depending on the deposition parameters. If the deposition parameters are such that a cluster collides with other adsorbed species before getting desorbed, it starts growing in size. After a certain critical size is reached, the cluster becomes thermodynamically stable and the nucleation barrier is said to have been overcome. This step involving the formation of stable, chemisorbed, critical-sized nuclei is called the *nucleation* stage.
4. The critical nuclei grow in number as well as in size until a saturation nucleation density is reached. The nucleation density and the average nucleus size depend on a number of parameters such as

the energy of the impinging species, the rate of impingement, the activation energies of adsorption, desorption, and thermal diffusion, and the temperature, topography, and chemical nature of the substrate. A nucleus can grow both parallel to the substrate by surface diffusion of the adsorbed species, as well as perpendicular to it by direct impingement of the incident species. In general, however, the rate of lateral growth at this stage is much higher than the perpendicular growth. The grown nuclei are called *islands*.

5. The next stage in the process of film formation is the coalescence stage, in which the small islands start coalescing with each other in an attempt to reduce the surface area. This tendency to form bigger islands is termed *agglomeration* and is enhanced by increasing the surface mobility of the adsorbed species, as, for example, by increasing the substrate temperature. In some cases, formation of new nuclei may occur on the areas freshly exposed as a consequence of coalescence.

6. Larger islands grow together, leaving channels and holes of uncovered substrate. The structure of the films at this stage changes from discontinuous island type to porous network type. A completely continuous film is formed by filling of the channels and holes.

The growth process may be summarized as consisting of a statistical process of nucleation, surface-diffusion controlled growth of the three-dimensional nuclei, and formation of a network structure and its subsequent filling to give a continuous film. Depending on the thermodynamic parameters of the deposit and the substrate surface, the initial nucleation and growth stages may be described as of (a) layer type, (b) island type, and (c) mixed type (called Stranski–Krastanov type). This is illustrated in Fig. 1.1. In almost all practical cases, the growth takes place by island

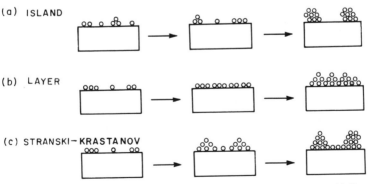

Figure 1.1. Basic growth processes: (a) island, (b) layer-by-layer, and (c) Stranski–Krastanov type.

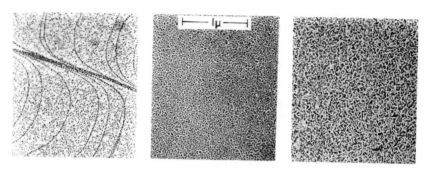

Figure 1.2. Transmission electron micrographs of 15, 45, and 75 Å thick argon-sputtered Au films deposited on NaCl at 25°C at a rate of ~1 Å/sec (K. L. Chopra, unpublished).

formation. The subsequent growth stages for a Au film sputter-deposited on NaCl at 25°C, as observed in the electron microscope, are shown in Fig. 1.2.

Except under special conditions, the crystallographic orientation and the topographical details of different islands are randomly distributed, so that when they touch each other during growth, grain boundaries and various point and line defects are incorporated into the film due to mismatch of geometrical configurations and crystallographic orientations, as shown in Fig. 1.3. If the grains are randomly oriented, the films show a ring-type diffraction pattern and are said to be polycrystalline. However, if the grain size is small ($\gtrsim 20$ Å), the films show halo-type diffraction patterns similar to that exhibited by highly disordered or amorphous (noncrystalline) structures. It is to be noted that even if the orientation of different islands is the same throughout, as obtained under special deposition conditions (discussed later) on suitable single-crystal substrates, a single-crystal film is not obtained. Instead, the film consists of single-crystal grains oriented parallel to each other and connected by low-angle grain boundaries. These films show diffraction patterns similar to those of single-crystals and are called epitaxial/single-crystal films.

Besides grain boundaries, epitaxial films may also contain other structural defects such as dislocation lines, stacking faults, microtwins and twin boundaries, multiple-positioning boundaries, and minor defects arising from aggregation of point defects (for example, dislocation loops, stacking faults, and tetrahedra and small dotlike defects). Note that defects such as stacking faults and twin boundaries occur much less frequently in polycrystalline films. Dislocations of density $\sim 10^{10}$ to 10^{11} lines cm^{-2} are the most frequently encountered defects in polycrystalline films and are largely incorporated during the network and hole stages, due to displacement (or orientation) misfits between different islands. Some other mechanisms

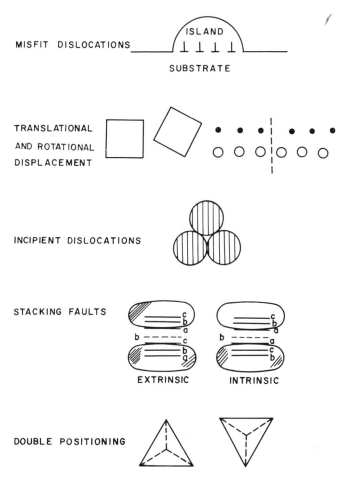

Figure 1.3. A schematic diagram showing incorporation of defects in a thin film during growth.

which may give rise to dislocations in thin films are: (1) substrate–film lattice misfit, (2) presence of inherent large stresses in thin films, and (3) continuation of the dislocations ending on the substrate surface into the film.

After a continuous film is formed, the anisotropic growth takes place normal to the substrate in the form of cylindrical columns. The lateral grain size (or the crystallite size) of a film is primarily determined by the initial nucleation density. If, however, recrystallization takes place during the coalescence stage, the lateral grain size is larger than the average separation of the initial nuclei, and the average number of grains per unit area of the film is less than the initial nucleation density. The grain size normal to the substrate is essentially equal to the film thickness for small ($\gtrsim 1\ \mu$m) thicknesses. For thicker films, renucleation takes place at the surface of previously grown grains, and each vertical column grows multigranularly with possible deviations from normal growth.

1.2.1. Structural Consequences of the Growth Process

The microstructural and topographical details of a thin film of given material depend on the kinetics of growth and hence on the substrate temperature, source (of impinging species) temperature, chemical nature, and topography of the substrate and gas ambients. These parameters influence the surface mobility of the adsorbed species, kinetic energy of the incident species, deposition rate, supersaturation (i.e., the value of the vapor pressure/solution concentration above that required for condensation into the solid phase under thermodynamical equilibrium conditions), the condensation or sticking coefficient (i.e., the fraction of the total impinging species adsorbed on the substrate), and the level of impurities. Let us now see how the physical structure is affected by these parameters.

1.2.1.1. Microstructure

The lateral grain size is expected to increase with decreasing supersaturation and increasing surface mobility of the adsorbed species. As a result, deposits with well-defined large grains are formed at high substrate and source temperatures, both of which result in high surface mobility. The transmission electron micrographs of 100-Å thick Au films deposited on NaCl at 100, 200, and 300°C by vacuum evaporation illustrate (Fig. 1.4) the effect of substrate temperature. Note that increasing the kinetic energy of the incident species, as, for example, by increasing the source temperature in the case of deposition by vacuum evaporation and by increasing the sputtering voltage in the case of deposition by sputtering, also increases

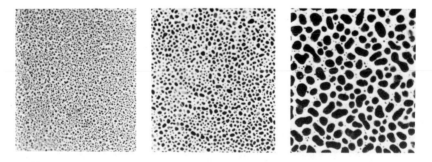

Figure 1.4. Transmission electron micrographs of 100 Å thick Au films vacuum evaporated onto NaCl at 100, 200, and 300°C (K. L. Chopra, unpublished).

the surface mobility. However, at sufficiently high kinetic energies, the surface mobility is reduced due to the penetration of the incident species into the substrate, resulting in a smaller grain size. This effect of the kinetic energy of the impinging species on grain size is more pronounced at high substrate temperatures. Also the effect of substrate temperature on grain size is more prominent for relatively thicker films.

The grain size may also be modified by giving the film a postdeposition annealing treatment at temperatures higher than the deposition temperature. The higher the annealing temperature, the higher is the grain size obtained. The effect of heat treatment is again more pronounced for relatively thicker films. It should be noted that the grain growth obtained during postdeposition annealing is significantly reduced from that obtained by depositing the film at annealing temperatures, because of the involvement of high-activation-energy process of thermal diffusion of the condensate atoms in the former case as compared to the process of condensation of mobile species in the latter.

For a given material–substrate combination and under a given set of deposition conditions, the grain size of the film increases as its thickness increases. However, beyond a certain thickness, the grain size remains constant, suggesting that coherent growth with the underlying grains does not go on forever and fresh grains are nucleated on top of the old ones above this thickness. This effect of increasing grain size with thickness is more prominent at high substrate temperatures. The effect of various deposition parameters on the grain size is summarized qualitatively in Fig. 1.5. It is clear that the grain size cannot be increased indefinitely because of the limitation on the surface mobility of the adsorbed species.

The formation of large-grain-sized epitaxial/single-crystal films under certain conditions has been mentioned earlier. The conditions favoring epitaxial growth are: high surface mobility as obtained at high substrate temperatures, low supersaturation, clean, smooth, and inert substrate surfaces, and crystallographic compatibility between the substrate and the deposit material. Films in which only a particular crystallographic axis is oriented along a fixed direction (due to preferential growth rate) are called *oriented* films. In contrast to epitaxial films which require a suitable single-crystal substrate, oriented films may also be formed on amorphous substrates. On the other extreme of thin film microstructures, highly disordered, very fine-grained, noncrystalline deposits with grain size $\gtrsim 20$ Å and showing halo-type diffraction patterns similar to those of amorphous structures (i.e., having no translational periodicity over several interatomic spacings) are obtained under conditions of high supersaturation and low surface mobility. The surface mobility of the adsorbed species may be inhibited, for example, by decreasing the substrate temperature, by introducing reactive impurities into the film during growth, or by codeposition of materials

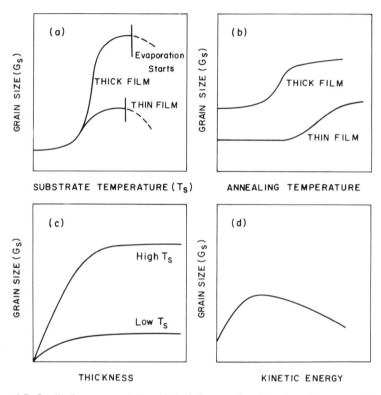

Figure 1.5. Qualitative representation of the influence of various deposition parameters on the grain size of thin films.

of different atomic sizes and low surface mobilities. Under these conditions, the film is amorphouslike and grows layer by layer.

1.2.1.2. Surface Roughness

Under conditions of low nucleation barrier and high supersaturation, the initial nucleation density is high and the size of the critical nucleus is small. This results in fine-grained, smooth deposits which become continuous at small thicknesses. On the other hand, when the nucleation barrier is large and the supersaturation is low, large but few nuclei are formed, as a result of which coarse-grained rough films, which become continuous at relatively large thicknesses, are obtained. High surface mobility, in general, increases the surface smoothness of the films by filling in the concavities, except in special cases where the deposit material has a tendency to grow preferentially along certain crystal faces because of either large anisotropy in the surface energy or the presence of faceted roughness on the substrate.

A further enhancement in surface roughness occurs if the impinging species are incident at oblique angles instead of falling normally on the substrate. This occurs largely due to the shadowing effect of the neighboring columns oriented toward the direction of the incident species. Figure 1.6 shows the topography of two rough film surfaces, one (a) obtained by oblique deposition and the other (b) obtained by etching of a columnar structure. Also shown in the figure is the topography of a smooth and a rough CdS film prepared by controlled homogeneous precipitation[13] under different conditions.

A quantitative measure of roughness, the roughness factor, is the ratio of the real effective area to the geometrical area. The variation of the

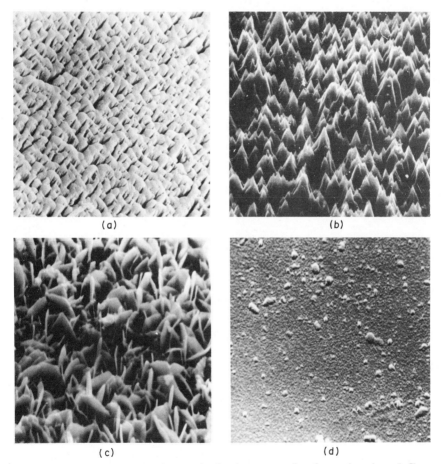

(a)   (b)

(c)   (d)

Figure 1.6. Scanning electron micrographs showing topography of smooth and rough films: (a) obliquely deposited GeSe film; (b) etched CdS film (vacuum evaporated); (c) rough CdS film (solution grown); (d) smooth CdS film (solution grown).

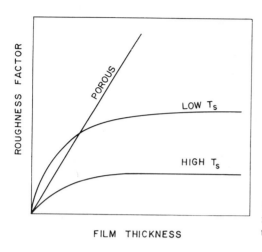

FILM THICKNESS

Figure 1.7. Qualitative variation of roughness factor as a function of film thickness. T_s: substrate temperature.

roughness factor with thickness for a number of cases is qualitatively illustrated in Fig. 1.7. In the case of porous films, the effective surface area can be hundreds of times the geometrical area.

1.2.1.3. Density

Density is an important parameter of physical structure. It must be known for the determination of the film thickness by gravimetric methods. A general behavior observed in thin films is a decrease in the density with decreasing film thickness. This is qualitatively illustrated in Fig. 1.8. Discrepancies observed in the value of the thickness at which the density of a given film approaches its bulk value are attributed to differences in the

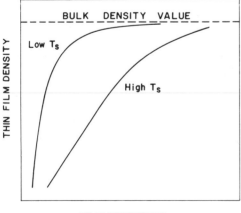

FILM THICKNESS

Figure 1.8. Qualitative variation of density as a function of film thickness.

deposition conditions and measurement techniques employed by different observers.

In the case of porous films, which are formed due to incorporation of gaseous impurities under conditions of poor vacuum and high supersaturation, the density can be as low as 2 to 3% of the bulk density, even in thick films.

1.2.1.4. Adhesion

The adhesion of a film to the substrate is strongly dependent on the chemical nature, cleanliness, and the microscopic topography of the substrate surface. The adhesion of the films is better for higher values of (1) kinetic energy of the incident species, (2) adsorption energy of the deposit, and (3) initial nucleation density. Presence of contaminants on the substrate surface may increase or decrease the adhesion depending on whether the adsorption energy is increased or decreased, respectively. Also the adhesion of a film can be improved by providing more nucleation centers on the substrate, as by using a fine-grained substrate or a substrate precoated with suitable materials. Loose and porous deposits formed under conditions of high supersaturation and poor vacuum are less adherent than the compact deposits.

1.2.1.5. Metastable Structures

In general, departures from bulk values of lattice constants are found only in ultrathin films. The lattice constants may increase or decrease, depending on whether the surface energy is negative or positive, respectively. As the thickness of the film increases, the lattice constants approach the corresponding bulk values.

A large number of materials when prepared in thin film form exhibit new metastable structures not found in the corresponding bulk materials. These new structures may be either purely due to deposition conditions or may be impurity/substrate stabilized. Some general observations regarding these new structures in thin films are as follows: (1) Most of the materials, in pure form or in combination with appropriate impurities, can be prepared in amorphous form. (2) The distorted NaCl-structure bulk materials tend to transform to the undistorted form in thin films. (3) The wurtzite compounds can be prepared in sphalerite form and vice versa. (4) Body centered cubic (bcc) and hexagonal close packed (hcp) structures have a tendency to transform to face centered cubic (fcc) structure. Some common examples of such abnormal stuctures found in thin films are: amorphous Si, Ge, Se, Te, and As; fcc Mo, Ta, W, Co, and β-Ta, etc. (all due to deposition conditions); and fcc Cr/Ni, bcc Fe/Cu, and fcc Co/Cu (all due to the

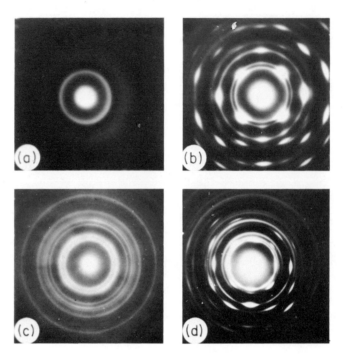

Figure 1.9. Electron-diffraction patterns of ~500 Å thick Zr films, ion-beam sputtered onto NaCl at (a) 23°C, amorphous; (b) 250°C, fcc; (c) 450°C, hcp; and (d) fcc annealed at 675°C in vacuum, fcc + hcp (after Ref. 14).

influence of the substrate). Note that these abnormal metastable structures transform to the stable normal structures on annealing. This is illustrated[14] in Fig. 1.9 for the case of ion-beam-sputtered Zr films.

1.2.2. Solubility Relaxation

Another consequence of the thin film growth process is the phenomenon of solubility relaxation. The atomistic process of growth during codeposition allows doping and alloying of films. Since thin films are formed from individual atomic, molecular, or ionic species which have no solubility restrictions in the vapor phase, the solubility conditions between different materials on codeposition are considerably relaxed. This allows the preparation of multicomponent materials, such as alloys and compounds, over an extended range of compositions as compared to the corresponding bulk materials. It is thus possible to tailor-make materials with desired properties, which adds a new and exciting dimension to materials technology. An important example of this technology of tailor-making materials is the formation of hydrogenated amorphous Si films for use in solar cells. Hydrogenation has made it possible to vary the optical band gap of amorphous Si from 1 eV to about 2 eV and to decrease the density of dangling bond states in the band gap so that doping (n and p) is made possible.

We will now introduce the reader to the technology of thin films. The following sections describe the commonly used techniques for deposition of thin films.

1.3 Vapor Deposition Techniques

Vapor deposition techniques can be broadly divided into two categories: (1) physical vapor deposition, and (2) chemical vapor deposition. Although genuine differences exist between the two classes, the line of demarcation is not really sharp and in many cases the techniques involve features of both classes. Nevertheless, the two are discussed separately. Only brief accounts of the important techniques will be given here. For detailed accounts, reference should be made to the standard textbooks[7-12] on the subject.

1.3.1. Physical Vapor Deposition (PVD)

1.3.1.1. General Remarks

The most important characteristic feature of PVD techniques is that the transport of vapors from the source to the substrate takes place by physical means. This is achieved by carrying out the deposition essentially in a vacuum of such magnitude that the mean free path (mfp) of the ambient gas molecules is greater than the dimensions of the deposition chamber and the source-to-substrate distance. Under such low-pressure ambient conditions, the transport of the material from the source to the substrate occurs by molecular beams.

The vapor species of a solid material may be created either by thermal evaporation, or by mechanically knocking out the atoms or molecules, from the surface by using energetic heavy particles. The process of deposition in the former case is referred to as vacuum evaporation and in the latter as sputtering. In pure PVD techniques, the deposits are formed from atomic or molecular units, simply by the physical process of condensation, for example, deposition of Cu films from Cu vapors, and that of SiO from SiO vapors. It should be noted, however, that some apparently simple-looking vaporization–condensation processes may actually involve a chemical reaction. For example, when ZnS is evaporated for preparing ZnS films, it dissociates into Zn and S, and ZnS films are formed by condensation of this two-component vapor system via a chemical reaction (Zn + S → ZnS) on the substrate. In some cases, a chemical reaction between different vapors is deliberately made to occur at the substrate. For example, thin films of multicomponent materials, such as alloys, compounds, and metal–dielectric mixtures (cermets) which are difficult to vaporize either because of very high vaporization temperature or because of problems associated with dissociation/decomposition, can be prepared by simultaneous condensation of the vapors of the individual components produced from separately controlled sources. This type of PVD process is known as *multisource*, or *coevaporation/cosputtering*. Thin films of oxides, nitrides, carbides, and hydrides of a number of metals can be prepared by evaporation/sputtering in a suitable controlled ambient, for example, O_2 (for oxides), N_2/NH_3 (for nitrides), C_2H_4 (for carbides), and H_2 (for hydrides). This type of PVD process is termed as *reactive evaporation/reactive sputtering*.

We thus see that in PVD techniques both vapor transport as well as deposition occur by physical processes. In some cases, however, a simple chemical reaction involving direct elemental combination may occur at the substrate during deposition. This particular feature of deposition via a chemical reaction is a characteristic of the chemical vapor deposition (CVD) techniques. The distinction between PVD and CVD techniques is thus ill defined. It should be pointed out, however, that whereas the chemical reactions involved in PVD are no more complex than direct elemental combination, those encountered in CVD are always far more complex.

1.3.1.2. Vacuum Evaporation

Vacuum evaporation is one of the most widely used deposition techniques. As the name implies, the technique consists of vaporization of the solid material by heating it to sufficiently high temperatures and condensing it onto a cooler substrate to form a film. Heating of the material can be carried out directly or indirectly (via a support) by a variety of methods. The simplest and the most common method is to support the material in a filament-basket (Fig. 1.10a) or boat (Fig. 1.10d), which is heated electrically. The material may also be supported directly on a wire in some cases, for example, by wrapping a thin foil of the material around the wire. The use of the basket and the wire type of sources is possible only in those cases where the deposit material either sublimes or wets the support in melted form so that it is prevented from falling due to surface tension. Evaporation sources are generally made of refractory materials such as W, Mo, Ta, and Nb, with or without a ceramic coating. Crucibles of insulating materials such as quartz, graphite, alumina, berrylia, and zirconia are heated indirectly by supporting them in a metal cradle. The choice of the filament

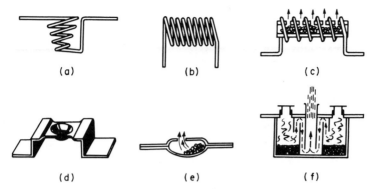

Figure 1.10. Some common thermal-evaporation sources: (a) basket, (b) spiral, (c) multiple-vapor-beam source consisting of a quartz tube with several holes ~1 mm diameter and heated by a tungsten spiral (Chopra, unpublished), (d) dimpled boat, (e) asymmetric oven-type point source, and (f) baffled chimney (also called Drumheller source).

or boat material is primarily determined by the evaporation temperature and the resistance to alloying and/or chemical reaction with the evaporant. Vapor sources of various types, geometries, and sizes can be easily constructed or obtained commercially. The geometries of some other sources are also shown in Fig. 1.10. Direct heating of the evaporant material can be done electrically by passing current through it, or by focusing an electron or laser beam on it.

The evaporated species, in the case of elements, consist of neutral single atoms except for S, Se, Te, Bi, P, Sb and As which vaporize in the form of polyatomic clusters. In the case of alloys and compounds, vaporization is generally accompanied by dissociation/decomposition because of differences in the vapor pressures of the various constituents or because of thermal instability. However, if the constituents are equally volatile, congruent evaporation occurs. The compositions of the vapor and the condensate differ from that of the source if evaporation is not congruent. This difference may be further aggravated if the condensation coefficients of the various constituent vapor atoms or molecules differ from each other. The tendency to dissociate is greater at high evaporation temperatures and low ambient pressures. Very few compounds, such as MgF_2, B_2O_3, CaF_2, SiO, GeO, and SnO, evaporate directly without dissociation.

According to the Langmuir–Dushman theory of the kinetics of evaporation, the rate of free evaporation of atoms or molecules from a clean surface of unit area in vacuum is given by

$$N_e = 3.513 \times 10^{22} p_e (1/MT)^{1/2} \text{ molecules cm}^{-2}\text{s}^{-1} \qquad (1.1)$$

where p_e is the equilibrium vapor pressure (in Torr; 1 Torr = 1 mm Hg) of the evaporant under saturated vapor conditions at a temperature T, and

M is the molecular weight of the vapor species. The rate of condensation of the vapors (or the deposition rate) depends not only on the evaporation rate but also on the source geometry, its position relative to the substrate, and the condensation coefficient.

Because of collisions with the ambient gas molecules, a fraction of the vapors, proportional to $\exp(-d/\lambda)$, is scattered and hence randomized in direction within a distance d during their transfer through the gas. The mean free path λ for air at 25°C and pressures of 10^{-4} and 10^{-6} Torr is about 45 and 4500 cm, respectively. Thus, pressures lower than 10^{-5} Torr are necessary to ensure a straight-line path for most of the evaporated species and for substrate-to-source distance of $\sim$10 to 50 cm in a vacuum chamber. Good vacuum is also necessary for producing contamination-free deposits.

A parameter of interest in understanding the influence of the degree of vacuum on the purity of films is the impingement rate of the ambient gas molecules. Besides the evaporant vapor species, the substrate is also impinged by the ambient gas molecules at a rate given by an expression similar to Eq. (1.1), with parameters N_g, P_g, T_g, and M_g referring to gas molecules. Table 1.1 lists the mfp and impingement rate of air molecules at different pressures. Note that at constant gas and evaporation temperatures, the ratio N_g/N_e is proportional to p_g/p_e. Values of N_g in Table 1.1 show that under the commonly employed experimental conditions of vacuum ($\sim$$10^{-5}$ Torr) and deposition rates ($\sim$1 Å sec^{-1}), the impingement rate of the gas molecules is relatively quite large, so that if the sticking coefficient of the gas molecules is not very small, a significant amount of gas sorption could occur. Fortunately, however, the sticking coefficient of the gas molecules at elevated temperatures is negligibly small, thereby making the pressures ($\sim$$10^{-6}$ Torr) good enough for deposition of clean

Table 1.1. Some Data on the Residual Air at 25°C in a Typical Vacuum Used for Film Deposition (After Ref. 8)

Pressure (Torr)	Mean free path (cm)	Collisions/sec (between molecules)	Molecules/ cm^2 sec (striking surface)	Monolayers/ sec^a
10^{-2}	0.5	9×10^4	3.8×10^{18}	4400
10^{-4}	51	900	3.8×10^{16}	44
10^{-5}	510	90	3.8×10^{15}	4.4
10^{-7}	5.1×10^4	0.9	3.8×10^{13}	4.4×10^{-2}
10^{-9}	5.1×10^6	9×10^{-3}	3.8×10^{11}	4.4×10^{-4}

^{a} Assuming that the condensation coefficient is unity.

films, except for those that are readily oxidizable, in which case relatively better vacuum conditions would be required.

The spatial distribution of the vapors from different types of sources such as point, wire, small surface, extended strip, cylinder, and ring has been calculated and discussed thoroughly by Holland.[7] For the ideal case of deposition from a clean, uniformly emitting point source onto a plane substrate, the rate of deposition varies as $\cos \theta / r^2$ (Knudsen's cosine law), where r is the radial distance of the substrate from the source and θ is the angle between the radial vector and the normal to the substrate. If t_0 is the thickness of the deposit where the vapors fall normally and t is the thickness at a distance x from this point, then the deposit distribution (assuming the same condensation coefficient throughout) is given by

$$\frac{t}{t_0} = \frac{1}{[1 + (x/h)^2]^{3/2}} \tag{1.2}$$

where h is the normal distance of the point source from the substrate. For evaporation from a small area onto a parallel plane substrate, the deposition rate is proportional to $\cos^2 \theta / r^2$, and the thickness distribution is given by

$$\frac{t}{t_0} = \frac{1}{[1 + (x/h)^2]^2} \tag{1.3}$$

In both the cases, the thickness decreases by about 10% for $x = h/4$. More complicated expressions for thickness distribution result for other types of sources.

We now briefly mention some important variants of the vacuum evaporation technique. One variation, known as flash evaporation, is particularly useful for deposition of multicomponent materials having widely different vapor pressures. In contrast to multisource evaporation, it uses only one source at a temperature sufficiently high to evaporate the less volatile material and does not require control of the vapor density or source temperature. In flash evaporation, small amounts of the material are continuously dropped into the heated source and evaporated discretely to completion. The net result of the simultaneous discrete evaporations is a vapor stream having uniform composition identical to that of the evaporant. In most cases of flash evaporation, since the supersaturation is high, the film composition is not much affected by the condensation coefficient.

Another very important variation of vacuum evaporation is molecular beam epitaxy (MBE).[15,16] As the name implies, MBE involves growth of epitaxial films by condensation of one or more controllably directed atomic or molecular beams each emerging from a point source in an ultrahigh vacuum (UHV) system. The low-density vapor beams are obtained from high-pressure Knudsen effusion sources which consist of a boron nitride crucible surrounded by heat shields and a liquid nitrogen shield. The vapors come out through a small orifice in the form of a highly directional beam.

Molecular beam epitaxy is basically a slow and controlled evaporation technique whose full potentials are realized only when it is used in conjunction with UHV analytical techniques (see Section 1.6) for obtaining *in situ* information on structure, topography, composition with its depth profile, and the chemical state of the surface of the film during growth. By carrying out the deposition onto ultraclean substrates at low rates ($\leqslant 1$ Å sec^{-1}), epitaxial growth of high perfection is achieved at relatively low substrate temperatures, as compared to conventional vacuum evaporation. In the case of multicomponent materials, the kinetics of condensation and thermodynamic reactions of various adsorbed species can be monitored layer by layer. Because of very slow deposition rates, MBE makes it possible to obtain multilayer structures of different materials in a predetermined sequence with layer thicknesses ranging anywhere from 10 Å to several microns, enabling the formation of quantum-well structures, heterojunctions, and graded composition/property structures. Further, three-dimensional geometrical structuring of thin films is also possible by using appropriate masks or shutters, as in the case of vacuum evaporation. Patterns may also be "written" directly using very fine ($\sim 10 \, \mu$m) vapor beams.

Molecular beam epitaxy is a very sophisticated and expensive process. It is particularly important for epitaxial growth studies of multicomponent materials, such as II–VI and III–V compound semiconductors.

1.3.1.3. Sputtering

Besides thermal evaporation, vapor species may also be created by mechanically knocking out the atoms or molecules from the surface of a solid material by bombarding it with energetic, nonreactive ions. The ejection process, known as *sputtering*, occurs as a result of momentum transfer between the impinging ions and the atoms of the target being bombarded. The sputtered species can be condensed on a substrate to form a thin film. The characteristics of the sputtering process as a deposition technique can be briefly summarized as follows: (1) The sputtered species, in general, are predominantly neutral. (2) The sputtering yield, defined as the number of ejected species per incident ion, increases with the energy and mass of the bombarding ions. The variation with energy shows a linear behavior in a small region above a threshold value determined by the sublimation energy of the target material. For higher energies, the yield approaches saturation, which occurs at higher energies for heavier bombarding particles. For example, Xe$^+$ bombardment of shows saturation above 100 KeV, whereas the Ar^+ bombardment curve is saturated at less than 20 KeV. However, at very high energies of the bombarding ions, the yield decreases because of the increasing penetration depth and hence increasing energy losses below the surface, with the consequence that not all the affected atoms are able to reach the surface to escape. (3) The yield increases as $(\cos \theta)^{-1}$ with increasing obliqueness (θ) of the incident ions. However, at large angles of incidence the surface penetration effect decreases the yield drastically. (4) The variation of the yield with the atomic number of the target shows an undulatory behavior with periodicity corresponding to one group of elements in the periodic table. (5) The yield for various elements bombarded with ions of a particular energy and mass varies by a maximum factor of 5 to 6. (6) The yield of a single crystal increases with decreasing transparency of the crystal in the direction of the ion beam. (7) The yield is rather insensitive to the target temperature except at very high temperatures where it shows an apparent rapid increase due to the accompanying thermal evaporation. (8) The energy of the ejected atoms shows a Maxwellian distribution with a long tail toward higher energies. With increasing bombarding energies, the peak of the distribution shifts only slightly toward higher energies because of the counteracting effect of the increased penetration of the more energetic ions. (9) The energies of the atoms or molecules sputtered at a given rate are about one order of magnitude higher than those thermally evaporated at the same rate. However, since sputtering yields are low and the ion currents are limited, sputter-deposition rates are invariably one to two orders of magnitude lower compared to thermal evaporation rates under normal conditions.

The sputtering process is very inefficient from the energy point of view, because most of the energy is converted to heat which becomes a serious limitation at high deposition rates. Provided the surface of a multicomponent target does not change metallurgically by thermal-diffusion, chemical reaction, or back-sputtering processes, the sputtering process ensures layer-by-layer ejection and hence a homogeneous film of composition corresponding to that of the target. This may not be so in the case of low-melting-point and high-yield materials. In the case of multiple targets, the composition is determined by the respective areas and yields of the target materials. The high energy of the ejected species and the attendant bombardment of the growing film (acting as anode) by electrons and negative ions have considerable influence on the growth of films and, in particular, yields highly adherent films.

In spite of being energy intensive, the sputtering process is best suited for depositing adherent films of multicomponent materials of any kind. Ions for sputtering may be produced either by establishing a glow discharge between the target and the substrate holder, or by using a separate ion-beam source. Accordingly, a sputter-deposition technique is referred to as glow-discharge sputtering or ion beam sputtering. Depending on the geometry of the target–substrate system and the mode of ion transport, a large number of sputtering variants have been developed. These are briefly described in the following sections.

    1.3.1.3a. *Glow-Discharge Sputtering.* A cheap and simple means of producing ions for sputtering is provided by the well-known phenomenon of glow discharge, which occurs when an electric field is applied between two electrodes in a gas at low pressure ($\sim 10^{-2}$ Torr). The gas breaks down to conduct electricity, above a certain voltage applied between the electrodes. The cathode (or Crooke's) dark space, across which most of the applied voltage drops, is the most important region for sputtering. Ions and electrons created at breakdown are accelerated across this region. The energetic positive gas ions strike the cathode to produce sputtering and cause emission of secondary electrons which are essential for sustaining the glow discharge. The accelerated electrons produce more ions by collision with gas atoms in the negative-glow region lying adjacent to the cathode dark space.

    Effective sputtering of the cathode target is possible only when both the number and the energy of the bombarding ions are large. The energy of the bombarding ions depends on the accelerating cathode fall (i.e., the voltage across the cathode dark space) and the thickness d of the cathode dark space, which is inversely proportional to the gas pressure p, that is, pd = const (Paschen's law). The number of ions striking the cathode (or the cathode current) depends on the gas pressure and the applied voltage. Initially, as the gas pressure is increased, the cathode current increases along with an increase in the area of the cathode glow, with the result that current density and hence the cathode fall remain constant. The discharge in this pressure region is said to be normal. As the cathode current is increased beyond the value for which the cathode is completely covered with cathode glow, the cathode current density and the cathode fall start increasing with increasing cathode current. In this region where both the current and the voltage increase together, the glow discharge is said to be abnormal. It is this abnormal glow discharge which is of interest to sputtering because, in this region, both the number as well as the energy of the ions is large. Increasing the gas pressure further increases the number of ions, but decreases their energy because the voltage across the cathode dark space falls due to thermionic emission from the cathode at high current densities. Also, at higher gas pressures, the sputtered atoms are prevented from reaching the substrate at the anode because of randomization due to the large number of collisions with the gas molecules. Thus, effective glow-discharge sputtering can take place only within an optimum pressure range of 20 to 100 mTorr.

Since the sputtered species are diffusely scattered by ambient gas molecules during their transit, they reach the substrate in randomized directions and energies. As a result of the diffuse nature of material transport, the atoms deposit at places not necessarily in the line of sight of the cathode. Also, note that because of the collisions the energetic ions hit the cathode at high oblique angles, which is actually helpful in increasing the yield. Glow-discharge sputterings can be performed using different electrode configurations as described below.

(i) *Diode Sputtering.* This is the simplest and the most widely used arrangement. Here, the target to be sputtered is made the cathode and the substrate is placed at the anode. The substrate may be kept at the anode potential or left floating. In the latter case, it will acquire a negative potential several volts relative to the anode and will thus attract impurities and gas ions leading to contamination of the film. Glow discharge is created by applying a dc voltage of 1–5 kV between the cathode and the anode separated by about 5 cm, with a current density ~ 1–10 mA cm^{-2}. The most commonly used gas for sputtering is Ar with $pd = 0.3$ Torr cm. The sputtering rate is proportional to the current for a constant voltage, which is a very convenient control parameter. Crude analysis of the distribution of the deposit in this parallel-plate diode configuration shows that, under optimum conditions of deposition, uniformity of the deposit extends to about half the area of the target when the cathode-to-anode distance is about twice the length of the cathode dark space.

Even if the sputtering system is initially pumped down to high vacuum and then sputtering gas of high purity is admitted, contaminants in the chamber may still appear from (1) outgassing as a result of plasma-discharge heating of the walls and components of the sputtering system, and (2) the glow-discharge decomposition of the oil vapors which may enter the chamber as a result of back-streaming from the diffusion pump operating at high pressures. Because of the continuous bombardment of the depositing film with neutral and negatively charged sputtering and contaminant ions, a large concentration (up to several percent depending on the deposition conditions) of gas and impurity atoms is trapped in the film in the case of diode sputtering.

Besides the more popular parallel-plate diode configuration, wire, cylindrical, and concave cathodes may be used for specific applications. The wire geometry gives deposition rates much higher than those in the parallel-plate configuration and is used for depositing on the inside of cylindrical substrates. The concave cathode produces annular-shaped deposits with very little deposit in the center. Multiple cathodes may be used for simultaneous or sequential sputtering for multicomponent or multilayer coatings.

(ii) Bias Sputtering. This arrangement differs from the previous one in that the substrate is biased at a large negative potential relative to the anode and is electrically insulated from the anode. In this case, the film is subjected to steady ion bombardment throughout its growth, which effectively cleans the film of adsorbed gases otherwise trapped in it as impurities. In another arrangement where sputter deposition is simultaneously accompanied by sputter cleaning by ion bombardment, an asymmetric ac, rather than dc, is applied between the cathode and the substrate placed at the anode, so that more material is deposited in one half-cycle than is removed by reverse sputtering in the other half-cycle. Bombardment removes not only the adsorbed gases but also the initial oxide layers which are responsible for good bonding of the film to the substrate. Good bonding may be retained by precoating with a positive bias on the substrate. The ac sputtering method is, however, more complicated and less efficient than bias sputtering.

(iii) Ion Plating. In this arrangement the deposition is actually obtained by vacuum evaporation from a filament. The substrate is kept at the cathode, so that the film is simultaneously sputter-cleaned by ion bombardment to give compact and adherent films. A more successful form of ion plating is to produce the ions of the deposit material by electron bombardment and condense them on an accelerating electrode to form a film.

(iv) Getter Sputtering. This process utilizes the gettering action of the sputtering material to purify the sputtering gas before it is actually used to sputter-deposit a film. In this arrangement, two cathodes of the material to be sputtered are symmetrically placed with respect to a Ni-anode can. The lower cathode is used to maximize the gettering action of active gases where they enter near the bottom of the can at ~100 mTorr pressure. After sputtering for a few minutes, the sputtered material from the second cathode is allowed to deposit on the substrate. The effectiveness of this method in reducing the reactive-gas content has been established.

The above-mentioned arrangements yield useful sputtering rates only in the 20–100 mTorr Ar pressure range, since the density of ions required for sputtering falls rapidly with decreasing pressure. The decreasing influence of gas atoms, the lower concentration of trapped impurities, the controlled direction, and the higher mean energy of the ejected species are some of the atractive features of low-pressure sputtering. Reasonably high sputtering rates can be obtained either by increasing the ionization efficiency of the available electrons, or by increasing the supply of ionizing electrons. The ionization efficiency of electrons can be increased either with a magnetic field, as in magnetron sputtering, or with an inductively coupled RF field, as in RF sputtering.

(v) Magnetron Sputtering. One convenient method of increasing the ionization efficiency of electrons is by increasing their path length by

applying a transverse magnetic field normal to the electric field. In the case of a planar diode configuration, such a field concentrates discharge on one side thus reducing the uniformity of the deposit. This drawback may be overcome by using a cylindrical cathode and a magnetic field parallel to its axis. Much more uniform deposits on three-dimensional substrates may be obtained by placing the substrate inside a hollow cathode.

Magnetron sputtering makes it possible to utilize the cathode discharge power very efficiently (up to 60%) to generate high current densities (up to 50 mA cm^{-2}) at relatively low ($\sim$500–1000 V) voltages to yield deposition rates which are at least one order of magnitude higher than those in the nonmagnetron sputtering systems. At a power density of 30 W cm^{-2}, a deposition rate as high as 25 μm/min has been achieved for Cu. The high deposition rates coupled with the fact that the film is not subjected to plasma and electron bombardment makes magnetron sputtering a very attractive technique for large-area, low-temperature deposition.

(vi) Assisted or Triode Sputtering. In this configuration, sputtering rates are increased by supplying auxiliary electrons from a thermionically emitting filament. Both the total ionization and the ionization efficiencies are increased by accelerating the electrons by means of a third electrode and injecting them into the plasma. The assisted sputtering process may further be enhanced by the presence of a magnetic field inclined to the lines of force between the cathode and the anode. A significant advantage of this system is the fine control of the current density and hence the sputtering rate by the easily varied magnetic field. Sputter deposition rates as high as those obtained in magnetron sputtering are obtained here.

(vii) RF Sputtering. Sputtering at low pressures ($\sim$10^{-3} Torr) is also possible by enhancing gas ionization with the help of a suitable RF field of several megahertz, applied directly to the anode through a capacitor (for metal sputtering) or via a high-frequency coil placed inside or outside the discharge region. An RF field may also be used additionally in any of the sputtering arrangements described already to obtain high sputtering rates at low pressures.

A very important application of RF sputtering is the sputtering of insulating materials, which is not possible using dc sputtering due to the buildup of positive charged sputtering gas ions which repel the bombarding (sputtering) ions. Methods devised to neutralize these surface charges by injecting electrons from a gun or by placing a metal screen over the cathode are not "clean." In RF sputtering, a high-frequency alternating potential is used to neutralize these surface charges periodically with plasma electrons which have a higher mobility than the positive ions. Typically, a 13.56-MHz power supply with $\sim$1–3 kW power and about 2 kV peak-to-peak voltage is used to couple the cathode through a matching network.

RF sputtering is a very versatile technique for high-rate deposition of semiconductors and insulators. In dc sputtering, deposition by reactive sputtering is limited to low rates because of the fact that a small amount of the reactive gas must be used to avoid the formation of an insulating layer at the cathode, which otherwise cannot be sputtered. However, since insulators can be RF-sputtered, the technique is ideally suited for high-rate reactive sputtering. By placing the filmed substrates on the cathode, RF sputtering may also be used to delineate desired patterns in it by sputter-etching. The absence of undercutting in contrast to chemical etching is the chief advantage of RF sputter-etching. The universality character is, however, a disadvantage since it does not allow selective or restricted etching of materials. This technique has been successfully used for etching of resistor patterns in cermet films with a definition significantly better than that obtained by conventional chemical etching techniques.

1.3.1.3b. *Ion-Beam Sputtering.* The sputtering systems described above are ineffective below 10^{-3} Torr because of the scarcity of ions. By producing ions in a high-pressure chamber and then extracting them into a differentially pumped vacuum chamber through suitable apertures with the help of suitable electron and ion optics, a high-density beam of ions may be obtained for sputtering in high vacuum.

The two ion sources commonly employed for sputtering are the duoplasmatron and that due to Kaufman. In the former, the ions are created in a glow or arc discharge chamber and are then extracted through apertures into a second chamber at a much lower pressure ($\sim 10^{-4}$–10^{-5} Torr). The Kaufman source employs a chamber geometry and an applied magnetic field in such a way that the thermionically emitted electrons must travel long spiral paths to an anode cylinder located in the outer diameter of the discharge region. This results in high ionization efficiency as well as a uniform plasma. By applying a potential difference between a pair of grids with precisely aligned holes, the ions are extracted from the sheath around the grid holes and then accelerated by this potential difference. The grid optics focuses these ions into a well-collimated beam. The beam may be neutralized by injecting low-energy electrons from a hot filament on the target side of the grids. Fully neutralized Ar beams of up to 25 mm diameter and $50\,\text{mA}\,\text{cm}^{-2}$ current at 500–1000 eV have been obtained from the Kaufman source. This source is well suited to both etching (called ion-beam milling) as well as sputter-deposition of conducting and nonconducting materials. The sputter-etching and sputter–deposition rates depend on the material and the angle of incidence of the beam, as discussed earlier.

1.3.1.4. *PVD Setups*

The single most important requirement for a PVD technique is a vacuum system with a glass or stainless steel chamber (or bell jar). Diffusion pump systems backed by rotary pumps yielding a vacuum $\sim 10^{-7}$ Torr continue to be the workhorses of vacuum deposition technology, largely because of their modest price, simplicity, and high speed. The ultrahigh vacuum range $\sim 10^{-8}$ to 10^{-11} Torr is easily obtained in an all-metal system evacuated by a combination of pump systems such as a diffusion pump with a special oil (e.g., polyphenyl ether), a cryogenic baffle-cum-pump, or an ion pump, cryopump, or trubomolecular pump backed by a sorption pump and assisted by a sublimation pump. An important point to remember is that each vacuum system has its own characteristic residual gas environment.

The ultimate pressure and the partial pressure of the residual gases depend not only on the type of vacuum system but also on the desorption characteristics of the hardware of the vacuum system. Therefore, materials for substrates, evaporation sources, masks, connections, O-rings, mechanical jigs, shutters, etc., must be compatible with UHV technology from the point of view of degassing and chemical reaction with the vapor species being deposited. Ideally, the vacuum system should also have facilities for structural and compositional analysis (see Section 1.6) in order to prepare films of a desired composition and structure. Figure 1.11 shows a schematic diagram of such an ideal vacuum system.

1.3.2. *Chemical Vapor Deposition (CVD)*

The foremost characteristic of a CVD[17,18] technique is that it necessarily involves a heterogeneous chemical reaction at the surface of a substrate without requiring vacuum as an essential condition for deposition. Here, continously flowing carrier gases and/or chemical reactions induced

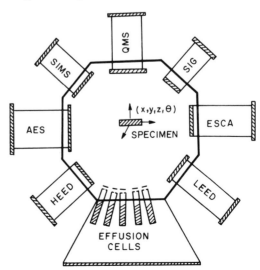

Figure 1.11. Schematic diagram of an ideal vacuum system having *in situ* analytical facilities.

by temperature or concentration gradients between the substrate and the surrounding gaseous ambient are responsible for the transfer of the vapor to the substrate. However, it may be pointed out that CVD may be carried out in pressures ranging anywhere from several atmospheres down to high vacuum, for reasons altogether different from those in PVD. Thus, CVD involves exposure of the substrate to the appropriate reactant vapors, and their reaction near or at the substrate surface to produce a film of the solid-phase reaction product. The deposition conditions should be such that the reaction occurs only at or near the substrate surface (heterogeneous reaction) and not in the gaseous phase (homogeneous reaction) to avoid formation of powdery deposits. The chemical reaction in CVD may be activated by the application of light, heat, RF field, x-ray radiation, electrical arc, glow discharge, electron bombardment, or by the catalytic activity of the substrate surface. The morphology of the deposited film is strongly influenced by the nature of the chemical reaction and the activation mechanism.

The major advantages of the CVD techniques are as follows: (1) In general, low-vacuum facilities are required and, thus, a relatively simple setup and fast recycling are possible; (2) high ($\sim 1\ \mu\mathrm{m/min}$) deposition rates are possible; (3) it is possible to deposit compounds with easily controlled stoichiometry; (4) it is relatively easy to dope the deposits with controlled amounts of impurities; (5) it is possible to deposit multicomponent alloys; (6) refractory materials can be deposited at relatively lower temperatures as compared to vacuum evaporation; (7) epitaxial layers of high perfection and low impurity content can be easily grown; (8) objects of complex shapes and geometries can be coated; and (9) *in situ* chemical vapor etching of the substrates prior to deposition is possible. However, the technique also has some drawbacks: (1) The generally complex thermodynamics and reaction kinetics are poorly understood; (2) higher substrate temperatures are required than those in the corresponding PVD techniques; (3) the reactive gases used for deposition and the volatile reaction products formed are, in most cases, highly toxic, explosive, or corrosive; (4) the corrosive vapors may attack the substrate, the deposited film, and the materials of the deposition setup; (5) the volatile products generated during the deposition process may lead to incorporation of impurities in the film; (6) the high substrate temperatures may lead to diffusion, alloying, or chemical reaction on the substrate surface, thus restricting the choice of the substrate; (7) high substrate temperatures may also give rise to segreggation effects when metastable multicomponent materials are being deposited; (8) it is difficult to control the uniformity of the deposit; and (9) masking of the substrate is generally difficult.

Basically, any chemical reaction between different reactive vapors which yields a solid-phase reaction product can be used for CVD. The substrate may, in some cases, take part in the reaction if the temperature is sufficiently high. For example, Si and Al substrates when exposed to O_2 atmospheres form a SiO_2 and Al_2O_3 film, respectively. The selection of a practically suitable reaction is dictated by several constraints. One has to take into account the fact that the actual course of the reaction may be much more complex and involve formation of intermediate species in accordance with the reaction kinetics, which depends on several factors such as flow rates, partial gas pressures, deposition temperature, temperature gradients, and nature of the substrate.

The chemical reactions utilized in CVD processes can be classified as: (1) decomposition reactions, (2) reduction reactions, (3) chemical transport reactions, and (4) polymerization.

(i) Decomposition. A general decomposition reaction can be written as

$$AB(gas) \rightleftharpoons A(solid) + B(gas)$$

Typical examples of the decomposition process activated by heat (pyrolytic decomposition) are

$$SiH_4 \xrightarrow{\text{800–1300°C}} Si + 2H_2$$

$$Ni(CO)_4 \xrightarrow{\text{200–300°C}} Ni + 4CO$$

$$CH_4 \xrightarrow{\text{1000–2000°C}} C + 2H_2$$

Whereas metal hydrides, metal carbonyls and complex carbonyls, metal borohydrides, and most of the organometallic compounds decompose at low temperatures ($<600°C$), metal halides, particularly the iodides, decompose at high temperatures. Many of the organic silicates decompose to yield SiO_2. Thin films of Al_2O_3 can be prepared by thermal decomposition of aluminum triethoxide at 550°C and aluminum tri-isopropoxide at 420°C.

For decomposition at low pressures, or with a large concentration of decomposition products, higher substrate temperatures may be required. Such an increase in substrate temperature is beneficial in improved crystallinity, purity, and adhesion of the film. Despite the simplicity of the decomposition process, difficulties may arise due to the formation of more than one nonvolatile reaction product, for example, carbon in the case of carbonyls and organometallic compounds, boron in borohydrides, and oxides in oxygen-containing compounds. A decomposition process may also be activated by electron bombardment or by glow discharge. A very important example of glow-discharge-activated decomposition is the preparation of amorphous Si thin films from SiH_4 for solar cell applications.

(ii) Reduction. A reduction reaction may be considered as a decomposition reaction aided by the presence of another vapor species, called the reductant. A reduction reaction occurs at temperatures much lower than the unaided decomposition reaction. Hydrogen or metal vapors are employed as reducing agents, while metal halides, carbonyl halides, oxyhalides, or other oxygen-bearing compounds are used to obtain the deposit material. In some cases, the addition of a reducing agent to the reactive vapor may also serve to prevent the codeposition of undesired oxides or carbides. The reducing action of the reductant increases in the sequence H_2, Cd, Zn, Mg, Na, K.

A typical example of CVD by reduction is the preparation of Si from the corresponding halide vapors using H_2 or Zn as the reducing agent, according to the reaction

$$SiCl_4 + 2H_2 \rightarrow Si + 4HCl$$

The use of H_2, which is not a very strong reductant, offers the advantage that it can be premixed with the halide vapors without causing a premature reaction in the gaseous phase which gives rise to powdery deposits.

Metals when used as a reductant may contaminate the deposit. To overcome this problem, the metal is used in stoichiometric proportions and the process is carried out at a reduced pressure. If in the reaction the reductant metal forms a halide less volatile than the parent metal (and may, therefore, be codeposited), the deposition conditions have to be maintained such that the pressure of the halide of the reductant metal is lower than its saturation pressure at the deposition temperature. For the reductant metals, the flourides, chlorides, bromides (except those of Zn), and iodides (except those of Zn and Mg) are less volatile than their parent metals, with fluorides being the least volatile and the iodides the most volatile. Further, alkali metal halides are the least volatile relative to the parent metal. Thus, iodides are preferable as deposition media, and alkali metals are least suitable as reductants.

(iii) Chemical Transport. The term chemical transport refers to the process of transferring of a nonvolatile deposit material to the substrate with the help of a highly volatile chemical vapor which converts it to a volatile compound by a chemical reaction. The volatile compound, formed as a consequence of this reaction, then decomposes on the substrate to yield a film of the source material. Chemical transport is achieved by shifting the reaction equilibrium to opposite directions at the source and the substrate either by maintaining different temperature and pressure conditions at the two locations [as in Eq. (a) below], or by introducing an additional chemical which reacts to form volatilized species to yield the

original deposit material at the substrate [as in Eq. (b) below].

$$Ti(solid) + 2NaCl(gas) \rightarrow TiCl_2(gas) + 2Na(gas) \qquad (a)$$

$$Si(solid) + SiO_2(solid) \rightarrow 2SiO(gas)$$
$$2SiO(gas) + O_2(gas) \rightarrow 2SiO_2(solid) \qquad (b)$$

Of the various reactions utilized for chemical transport, the disproportionation reactions, particularly the halide disproportionation, are the most widely used. Transport is accomplished by treating the nonvolatile metal with the vapor of its own higher-valent halide at a high temperature to yield a lower-valent volatile halide which, after transport into a cooler zone, disproportionates back into the higher-valent volatile halide and nonvolatile metal at the substrate. The higher-valent halide is recycled by feeding it back to the hotter zone. Thus the system can operate as a closed system. A typical example is the transport of Si in I_2 vapors.

$$Si(solid) + 2I_2(gas) \xrightarrow{\text{1100°C}} SiI_4(gas)$$

$$Si(solid) + SiI_4(gas) \xrightarrow{\text{1100°C}} 2SiI_2(gas)$$

$$2SiI_2(gas) \xrightarrow{\text{900°C}} Si(solid) + SiI_4(gas)$$

An additional advantage of transport reactions is that purification of the material also occurs in those cases (such as above) where the corresponding iodide vapors of the major contaminants of the source material have vapor pressures widely different from those of the main transporting species.

Transport reactions can also be used for gaseous etching, e.g., of Ge and Si with HCl, and of Al_2O_3 with flourinated hydrocarbons. It may also serve to remove contaminants and surface damage prior to deposition of epitaxial films.

(iv) Polymerization. As the name implies, polymerization involves linking together of monomeric molecules of organic and organic–inorganic composites. The polymerization process may be activated by (1) electron or ion bombardment, (2) irradiation with light, x-rays, or γ-rays, (3) electrical discharge, and (4) surface catalysis or surface recombination of monomers having free radicals. The films can be produced by (a) condensing the monomer vapors on the substrate and subjecting it simultaneously or sequentially to the activation mechanism, (b) activating polymerization in

the gas phase and allowing the polymerized product to deposit on the substrate, or (c) depositing the monomer film by other means and then activating the polymerization process.

The polymerized films have electrical properties ranging from semiconducting to insulating and have certain desirable characteristics such as complete surface coverage, good adhesion, low stress, and high plasticity.

In conclusion, among the important materials deposited by CVD techniques are semiconductors such as II–VI and III–V compounds, Si, and conducting doped oxides of Sn, In, and V, insulators such as SiO_2, Si_3N_4, BN, Al_2O_3, Ta_2O_5, TiO_2, Nb_2O_5, Al_2O_3, and AlN, and a number of metals.

The commonly used CVD systems cover a wide range from extremely simple laboratory setups to highly sophisticated, completely automated, electronically controlled and computerized industrial reactors. The systems can be either closed, permitting complete recovery of the reagent species and recycling, or open, requiring an external supply of source material and extraction of the reaction components. A typical CVD system used for the deposition of Si p–n junction structures is shown[19] in Fig. 1.12.

1.4. *Solution Deposition Techniques*

In these techniques, the unit species of the material to be deposited are dispersed in a liquid medium (generally aqueous) and are always present in ionic form. Solution deposition techniques, therefore, inherently involve chemical and/or electrochemical reactions for the formation of the deposit

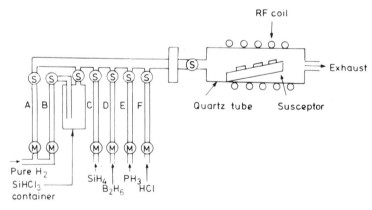

Figure 1.12. Schematic diagram of a CVD setup used for fabrication of Si p–n junction structures (S) Shut-off valves; (M) metering valves; (A, B, C, D, E, F) flowmeters (after Ref. 19).

material. Accordingly, the solution deposition techniques may be broadly classified into two categories: (1) chemical solution deposition (CSD) and (2) electrochemical deposition (ECD). Reactions such as precipitation, displacement, or reduction may be utilized in CSD techniques. ECD techniques, on the other hand, involve electrochemical reactions, i.e., chemical reactions necessarily involving interaction with an external source of electric current. In both cases, the substrate may or may not take part in the reaction. Two types of chemical reactions may be distinguished when the substrate plays an active role. In the first, called *conversion*, the substrate is converted to one of its compounds. In the second, called *displacement*, the substrate material is displaced by another from the solution.

Some general advantages of solution deposition techniques over vapor deposition techniques are as follows: (1) Experimental setups are much less sophisticated compared to those in vapor deposition techniques; (2) no expensive equipment such as vacuum systems is required for deposition; and (3) deposition is carried out at much lower temperatures ($<100°C$). However, the two types of solution deposition techniques suffer from the common drawback that preparation of ultraclean substrates using *in situ* techniques, such as plasma etching and ion bombardment as in the case of vapor deposition techniques, is not possible. Also the choice of the substrates is limited to only water-insoluble materials. The various solution deposition techniques are described briefly in the following sections.

1.4.1. *Chemical Solution Deposition (CSD)*

Chemical solution deposition techniques are generally immersion techniques in the sense that they involve simply dipping of the substrate into the reaction mixture for some time depending on the thickness required. The CSD techniques have the following advantages over the ECD techniques: (1) Since the question of nonuniform current density does not arise in CSD, deposits are formed more uniformly even on complex parts without any excessive buildup on projections and edges; (2) deposits can also be laid down directly on insulators; and (3) no power supplies and contacts are needed. According to the chemical reaction involved in deposition, CSD techniques can be classified as follows.

1.4.1.1. *Autocatalytic Reduction/Electroless Plating*

This technique is applicable for deposition of metal and metal-alloy films. In contrast to electrodeposition (Section 1.4.2.1) where the metal ions are converted to the corresponding elemental form (reduction) by taking electrons from an external source of current, in electroless plating a chemical reducing agent is used to supply the electrons for reduction.

The reduction, however, does not take place homogeneously throughout the volume of the solution, but only on certain catalytic surfaces. This means that, for deposition to continue, the metal being deposited must itself be catalytic in nature, and hence the name autocatalytic reduction. Deposition on noncatalytic surfaces (including insulators) requires sensitization of the surface, for example, by treating it successively in $SnCl_2$ and $PdCl_2$ solutions. This treatment gives rise to formation of catalytic Pd metal nuclei on the surface. The electroless deposition process can thus be described by the equation

$$M^{+n} + n\,e \text{ (supplied by the reducing agent)} \xrightarrow[\text{surface}]{\text{catalytic}} M^0$$

Several metals (Ni, Co, Cu, Pt, Pd, Au, and Ag) and their alloys have been successfully deposited by the electroless method using a number of chemical reducing agents such as hypophosphite, borohydride, formaldehyde, hydrazine, and ammine boranes and some of their derivatives. Silver is an exceptional case in that not only does the deposition take place indiscriminately on all types of surfaces in contact with the solution, but also the formation of Ag takes place homogeneously in the volume of the solution.

Due to incorporation of an element in the film from the reducing agent (for example, up to 13% P from hypophosphite and up to a few percent B from borohydride), special properties may be conferred on electroless deposits. The extreme hardness and the wear resistance properties of electroless Ni and the magnetic properties of electroless Co are some examples. It may, however, be pointed out that an electroless process·is more expensive compared to electrodeposition due to the high cost of the reducing agents. Also, the electroless baths are not very stable over long periods of time because of their tendency to develop catalytic particles in the solution. The electroless technique, therefore, must not be thought of as a substitute for electrodeposition and should be employed only in cases where unique deposit properties or process capabilities are desirable.

1.4.1.2. Homogeneous Precipitation and Solution Growth

The technique of controlled homogeneous precipitation is applicable to deposition of water-insoluble compounds and their solid solutions. In particular, II–VI and IV–VI compound semiconductors are of great importance. A number of compounds such as CdS, PbS, CdSe, PbSe, ZnS, ZnSe, SnO_2, and Bi_2S_3 have been successfully deposited.

For depositing thin films of a compound $M_m X_n$, a solution of M^{+n} ions with a complexing agent (or ligand) L (generally used ligands are NH_3, CN^-, EDTA, and trisodium citrate) added to it is prepared. Formation of complex ions $[M(L)_i]^{+n}$ is essential to control the reaction and avoid immediate precipitation of the compound in the solution when the precipitating anions are added to it. Complexation of the metal ions also avoids the precipitation of hydroxides in the solution, which is essentially alkaline, thus making film deposition possible. The precipitating agent is generally a compound which, upon hydrolysis, slowly generates the anions in the solution. For example, thiourea and thioacetamide generate S^{-2} ions and selenourea and sodium selenosulfate generate Se^{-2} ions. The cations are generated by decomposition of the complex ions according to the equation

$$[M(L)_i]^{+n} \to M^{+n} + iL$$

as for example, when the solution is heated. Compound formation starts when the ionic product $([M^{+n}]^m[X^{-m}]^n)$ exceeds the solubility product (i.e., the ionic product at saturation) and progresses slowly both on the substrates immersed as a film and in the solution as powder. It may be mentioned here that the solution must be vigorously stirred during deposition to encourage ion-by-ion deposition which gives rise to compact, pinhole-free, coherent, adherent, and specularly reflecting films and to inhibit the growth by adsorption of colloidal-compound particles formed in the solution, which gives rise to loose and powdery deposits. The quality of the films deposited further depends on the composition of the reaction bath. This is illustrated[13] in Fig. 1.13 for the case of CdS films.

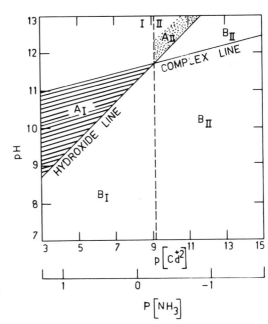

Figure 1.13. A plot of $[Cd^{+2}]$ as a function of $[NH_3]$ and $[OH^-]$ in the bath used for preparation of CdS films. Good quality films are obtained in regions A_I and A_{II}. Poor quality, powdery films are obtained in regions B_I and B_{II} (after Ref. 13).

This powerful technique has been used extensively[13,20-26] for deposition of II–VI and IV–VI compound semiconductor films. It has also been successfully exploited for the deposition of oxide films, for the first time in the authors' laboratory.[27] Deposition of oxide films is based on the fact that a large number of metal ions in alkaline solutions are precipitated as hydroxides or hydrous oxides, which upon heating to around 100–200°C yield the anhydrous oxides.

The technique offers the following advantages: (1) Films can be deposited on all kinds of hydrophilic substrates; (2) it is a very simple and inexpensive technique suitable for large-area deposition; (3) impurities in the initial chemicals can be made ineffective by suitably complexing them; and (4) in contrast to the difficulty of obtaining stoichiometric films by such techniques as reactive and coevaporation, reactive sputtering, and electrodeposition for deposition of compound films, deposits formed by homogeneous precipitation are always nearly stoichiometric. This is so because the basic building blocks are ions instead of atoms, so that the condition of charge neutrality maintains the stoichiometry. However, deviations from stoichiometry may arise if the metal ion is multivalent.

1.4.1.3. Spray Pyrolysis

A large number of metallic salt solutions when sprayed onto a hot substrate decompose to yield oxide films. Similarly films of sulfides and selenides can be prepared by pyrolytic decomposition of a solid complex compound formed on the surface of a substrate by spraying a mixed solution of the corresponding metallic salt and a sulfur/selenium-bearing compound. For example, $CdCl_2$ and $(NH_4)_2CS$ (thiourea) solutions when mixed at room temperature do not react to form CdS (provided the solution is not deliberately made alkaline). But when sprayed onto a substrate held at ~400°C, a solid complex of composition $CdCl_2 \cdot 2$ thiourea is formed, which decomposes on the substrate to yield CdS films.

A schematic block diagram of a typical spray-pyrolysis setup in operation in the author's laboratory is shown in Fig. 1.14. The atomization of the chemical solution into a spray of fine droplets is effected by the spray nozzle with the help of a filtered carrier gas which may (as in the case of SnO_x films) or may not (as in the case of CdS films) be involved in the pyrolytic reaction. The carrier gas and the solution are fed into the spray nozzle at predetermined and constant pressure and flow rates. The substrate temperature is maintained with the help of a feedback circuit which controls a primary and an auxiliary heater power supply. Large-area uniform coverage of the substrate is effected by scanning either or both the spray head and the substrate, employing mechanical or electromechanical arrangements.

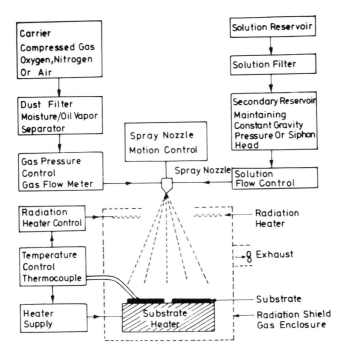

Figure 1.14. Schematic diagram of the experimental setup for spray pyrolysis used in the authors' laboratory.

The geometry of the gas and the liquid nozzles strongly determines the spray pattern, size distribution of droplets, and spray rate, which in turn determine the growth kinetics and hence the quality of the films obtained. These aspects of the spray pyrolysis technique are dealt with in detail in a recent review article by Chopra *et al.*[26]

Doped and mixed films can be prepared very easily, simply by adding to the spray solution a soluble salt of the desired dopant or impurity. Two very important examples of the materials deposited by this technique for device fabrication are (1) CdS for solar cells and (2) In F- or Sb-doped SnO_2 for conducting glass.

1.4.1.4. *Displacement Deposition, Chemiplating, or Immersion Plating*

If a substrate consisting of a less noble metal, say M_1, is immersed in a solution containing ions of a more noble metal, M_2, the more noble metal ions displace the less noble metal at the substrate surface, converting the latter into ions in the solution. A displacement reaction is, thus, described by

$$M_1^0 + M_2^{+n} \rightarrow M_2^0 + M_1^{+n}$$

For example, immersion of an iron rod into a $CuSO_4$ solution produces a Cu layer on the rod according to the reaction

$$Fe^0 + Cu^{+2} \rightarrow Cu^0 + Fe^{+2}$$

Displacement occurs as a result of establishment of a potential between the substrate surface and the solution. Deposition ceases when the substrate is completely covered unless the ions have a very high diffusion constant in the substrate. Except in some cases, for example, Cu on stainless steel for lubricating drawing wire and Sn on Al for protection of pistons from wear, deposits are generally loose, powdery, and nonadherent because of uncontrolled reaction.

One compound may also be converted to another by displacement. An important example of such a deposition is the formation[28] of Cu_2S film on CdS for solar cells by immersing it in a solution of Cu^+ ions. This technique of Cu_2S formation has also been referred to as chemiplating in the literature.

1.4.1.5. *Conversion Coatings*

Here the coating is formed by chemical treatment of a metallic substrate surface which is converted into one of its compounds. The coating forms an integral part of the substrate. The most important surface treatments are chromating and phosphating. Important metals, where the former treatment is particularly useful, are Zn, Cd, Cu, brass, Sn, Mg, and Al. Chromating consists of immersion of the metallic substrate in an aqueous solution of hexavalent Cr, such as CrO_4^{-2}. The metal is oxidized and the Cr(VI) is reduced to Cr_2O_3. The chromate coatings generally consist of oxides of the base metal, trivalent Cr (as Cr_2O_3) and hexavalent Cr [as chromate (CrO_4^{-2}) of the base metal]. The relative amounts of Cr(III) and Cr(VI) depend on the deposition conditions such as the bath temperature, pH, and nature of the base metal. The color of the coatings may vary from colorless to black, depending on the relative concentrations of Cr(III) and Cr(VI) and the thickness of the coating. Chromate conversion coatings are useful for protection against corrosion. They also provide a good base for subsequent application of paints.

Similar functions can also be served by phosphate conversion coatings, which are formed by immersing the substrate in phosphoric acid. Salts of

Mn and Zn may also be added to phosphoric acid. The most important phosphate coatings are those of Fe, Mn, and Zn. Phosphate coatings contain phosphates and oxides of the base metal.

1.4.2. Electrochemical Deposition (ECD)

As already mentioned, all ECD techniquues[29,30] require an external source of current for deposition. Obviously, they also require electrically conducting substrates. The various ECD techniques are briefly discussed in the following sections.

1.4.2.1. Electrodeposition

The occurrence of chemical changes due to the passage of electric current through an electrolyte (salt solution) is termed *electrolysis*, and the deposition of any substance on an electrode as a consequence of electrolysis is called *electrodeposition*.

The phenomenon of electrolysis is governed by the following two laws, first enunciated by Faraday in 1933: (1) The magnitude of the chemical change occurring is proportional to the electricity passed. (2) The masses of different species deposited at or dissolved from the electrodes by the same quantity of electricity are in proportion to their chemical equivalent weights.

When a metal M is dipped in a solution of its own ions, a dynamic equilibrium

$$M \rightleftharpoons M^{+n} + n\,e$$

is set up, as a consequence of which a resultant potential, called the *electrode potential*, develops between the electrode and the electrolyte, in the absence of an external voltage. The electrode thus gains a certain charge on its surface which attracts oppositely charged ions and water molecules, holding them at the electrode–electrolyte interface by electrostatic forces. Thus, the so-called electric double layer (the inner consisting of oriented water dipoles, and the outer, oppositely charged ions) is formed. During deposition, ions reach the electrode surface, move to stable positions on it while simultaneously releasing their ligands, release their charges, and undergo the stipulated electrochemical reaction. Fresh ions are supplied to the ion-depleted region near the electrode by either of the following means: (1) diffusion due to the concentration gradient, (2) migration due to the applied electric field, and (3) convection currents in the electrolyte. The growth of the deposits is determined by a number of parameters such as the current density, bath composition and temperature, electrode shape, nature of the counter-electrode, and mechanical agitation of the electrolyte. The main constituent of the bath is the electrolyte which serves to provide ions either in their simple form or in complex form. Additional chemicals such as wetting agents, brightening agents, alkalis, and acids may be added to the bath as required.

Almost all the metals, except for highly electronegative ones, can be electrodeposited from aqueous solutions. Alloys can be deposited by coelectrolysis, for which the electrode potentials should be brought very close to each other, as for example, by suitably complexing them and by reducing the concentration of the more noble metal ions.

Electrodeposition of compound semiconductors like CdSe, CdTe, Ag_2Se, and GaAs is also possible by codepositing the two components in elemental form by cathodic reduction of their corresponding ions and their subsequent reaction to form the corresponding compound. For example, deposition of CdTe film involves electrolytic codeposition of Cd and Te from Cd^{+2} and TeO_3^{-2} ions. Similarly, CdSe and GaAs are electrodeposited from Cd^{+2} and SeO_3^{-2} ions and Ga^{+3} and AsO_2^{-2} ions, respectively. The reactions involved at the cathode, for example, in the case of CdSe, are

$$Cd^{+2} + 2e \rightarrow Cd$$

$$SeO_3^{-2} + 3H_2O + 4e \rightarrow Se + 6OH^-$$

$$Cd + Se \rightarrow CdSe$$

Baranski and Fawcett[31] have used a new approach to deposit thin films of semiconducting metal chalcogenides such as CdS, HgS, PbS, Ti_2S, Bi_2S_3, Cu_2S, NiS, CoS, and CdSe on various conducting substrates. They used an electrolyte constituted of the corresponding metal salt and chalcogen in elemental form by dissolving the two in a suitable nonaqueous organic solvent such as DMF (dimethyl formamide), DMSO (dimethyl sulfoxide), or ethylene glycol. The nascent metal formed at the cathode rapidly reacts with the dissolved chalcogen to form the corresponding metal chalcogenide. The use of nonaqueous organic solvent avoids the problem of H_2 evolution at the cathode. Also, deposition is simpler because it involves electrolysis of a single electrolyte.

1.4.2.2. Anodization

Anodization[32] is an electroconversion process in which the working metal is made the anode in a suitable electrolyte solution so that, when an electric current is passed through the electrolyte, the anode surface is converted to one of its compounds. For example, anodization in acids and bases gives rise to oxide coatings, in thiourea it gives rise to sulfide coatings, and in NH_3 it gives rise to nitride coatings. The composition of the bath plays an important role in determining the quality of the films. For example, in the case of oxide coatings, if the bath is too acidic or too alkaline, film will dissolve during deposition and a porous structure will be obtained. With the optimum pH value, coherent oxide films of a number of metals such as Al, Ta, Nb, Si, Ti, and Zr can be obtained. The thickness of the oxide coating depends on the metal, the voltage applied, the temperature of the bath, and the time for which deposition is carried out. Other coatings formed by anodization are those of PbS, CdS, Bi_2S_3, and Si_3N_4.

1.4.2.3. Electrophoretic Deposition

Electrophoretic deposition[33] is a process by which electrically charged particles suspended in a liquid are deposited on an electrode by the application of an electric field. It is a very versatile technique, and practically any material can be deposited on a conducting substrate. The range of materials electrophoretically deposited includes metals, alloys, oxides, refractory compounds, polymers, and mixtures of various components. The first step in electrophoretic deposition involves the formation of a colloidal suspension of the material to be deposited. Colloidal particles are formed either by breaking up larger particles into smaller ones of colloidal size (10–5000 Å), or by effecting the merger of smaller particles, usually ions or molecules, to form larger aggregates of colloidal size. In the first process, colloidal-sized particles, say those produced by ball-milling, are suspended in an organic solvent such as acetone or methyl-, ethyl-, and propyl alcohol, where they spontaneously acquire electric charges. In the second process, a colloidal precipitate is formed *in situ* in a suitable medium using a chemical reaction. By this technique colloids of a large variety of materials including metals such as Au, Ag, and Cu and nonmetals such as S and Se, sulfides, hydroxides, and hydrous oxides can be conveniently prepared. The particles acquire electric charges due to ions which are adsorbed because of the large surface-to-volume ratio of the particles.

In either case, surfactants may be added to the colloid to change the charge of the particles. Sometimes it may also be necessary to add a small amount of a polymer for stabilizing the colloid. Particles as large as 20 μm may also be electrophoretically deposited. Deposits are formed on the cathode or the anode depending on the charge of the colloid. Codeposits can be formed from mixed suspensions, and the particles depositing on the electrode are usually in the same proportion as that existing in the suspension.

The technique has been successfully used for deposition of alumina on tungsten wires for insulation, ZnS for luminescent screens, CdS for solar cells, polytetrafluoroethylene (PTFE) for insulation of metallic wires, rubber for gloves, etc. The major advantages of this technique are as follows: (1) Uniform thickness is obtained even on complicated shapes; (2) very thick coatings can be deposited, and the thickness can be precisely controlled; (3) the deposition period is very short, generally of the order of a few seconds to minutes; and (4) no material is wasted.

However, in contrast to electrodeposition which involves deposition from ions in solution, resulting in compact and adherent coatings, electrophoretic deposits are usually loosely adherent coatings of powder requiring further treatment (such as annealing, sintering, and/or pressure compaction) before use to produce adherent, compact, and mechanically strong coatings.

1.4.2.4. Cathodic Conversion

This is another form of electroconversion developed by Nakayama *et al.*[34] to form Cu_2S films on CdS ceramic plates and extended in the author's laboratory[35] to deposition on CdS thin films. The deposition system consists of a thin film CdS cathode, a Cu anode, and a $CuSO_4$ electrolyte. The stoichiometry of the copper sulfide film formed, according to the reaction

$$CdS + 2Cu^{+2} + 2e \rightarrow Cu_2S + Cd^{+2} \text{ (goes into solution)}$$

depends on the deposition parameters such as the current density, the temperature and pH of the bath, and the concentration of the electrolyte. An advantage of this technique over other techniques of preparing Cu_2S is that unstable cuprous salts are avoided. Also the deposition time is considerably longer, thus affording better control of the stoichiometry.

1.5. *Thick Film Deposition Techniques*

Although thick films are not the subject matter of this book, the following thick film techniques are of considerable technical interest in the development of thin film devices and are therefore described briefly in the following sections.

1.5.1. *Liquid-Phase Epitaxy (LPE)*

Basically LPE involves the precipitation of a material from a cooling solution of the material in the melt of a suitable element, onto an underlying single-crystal substrate. The solution and the substrate are kept apart, and contact is achieved either by "tipping" the furnace with the solution or by dipping the substrate into the solution in a vertical furnace. The solution is saturated with the growth material at the desired growth temperature and is then allowed to cool in contact with the substrate surface at a rate and for a time interval appropriate for the generation of the desired layer. Under optimum conditions, the layer grows as an extension of the single-crystal substrate. The LPE technique has been very successfully applied to prepare device quality III–V compound films.

1.5.2. *Screen Printing*

The technology of screen printing offers a flexible, inexpensive method of preparation of pattern-delineated thick films for resistors, electrodes, and other components in circuits. Screen printing can also be used for deposition of compound semiconductor films for devices.

The procedure for screen printing consists of dispersing a paste (called the ink) of the material to be deposited on a mesh-type screen on which the desired pattern is photolithographically defined (see Section 1.7) such that open mesh areas in the screen correspond to the configuration to be printed. The substrate is placed a short distance beneath the screen. A flexible wiper, called the squeegee, then moves across the screen surface, deflecting the screen vertically, bringing it into contact with the substrate, and forcing the paste through the open mesh areas. On removal of the squeegee, the screen regains its original position, leaving behind the printed paste pattern on the substrate. The pattern consists of a series of discrete spots each corresponding to a mesh opening in the screen. The substrate is allowed to stand at ambient temperature for some time to enable the paste to coalesce to form a coherent level film.

Since, in addition to the deposit material, the paste contains other organic materials such as an organic suspension medium, a bonding agent (finely divided glass frit), and an organic diluent, the coating requires suitable heat treatment for removing these extra materials. The screen-printed film is first heated at temperatures ranging from 70–150°C for 15 to 30 min to dry the film and remove more volatile paste components. The large amount of organic material left in the paste after drying is removed at a relatively low temperature ($\sim$400°C) by carbonizing and oxidizing. The binder removal is carried out at the first low-temperature phase of the final firing process which is carried out at high temperatures ($\sim$1000°C). The glass component of the ink melts to form a vitreous medium which consolidates the printed layer and promotes adhesion to the substrate. The important properties of the film are determined by the chemical reactions taking place in the high-temperature zone of the furnace.

The substrates to be used for screen printing must have a uniform surface texture for good adhesion, ability to withstand high temperatures, a minimum of distortion or bowing, high mechanical strength, high thermal conductivity, chemical and physical compatibility with the material to be printed, and, lastly, low cost in quantity production. Ceramic materials such as alumina, berrylia, zirconia, magnesia, and glass are commonly used for substrates.

1.5.3. Melt Spinning

This is a very promising technique for producing rapidly quenched ribbons at high speeds of several meters per second. By bringing the molten material to be cast into ribbons in contact with a spinning wheel, the liquid drop is pulled out in the form of a ribbon. The rate of solidification of the melt as well as the ribbon thickness and width are dependent on the nozzle dimensions, the detailed shape of the molten drop in contact with the spinning wheel surface, the linear velocity of the wheel, and the heat transfer processes. At high spinning speeds, this technique has been used extensively to obtain metallic glass ribbons of 10–100 μm thickness formed at quenching rates $\sim 10^6$ K/sec. The use of this technique to obtain rapidly quenched, large-grain-sized ribbons of Si for solar cell applications offers promising possibilities. In principle, by employing multiple as well as sequential nozzles, wide Si ribbons coated on metallic substrates should be obtainable for solar cell applications.

1.5.4. Dip Coating, Spinning, and Solution Casting

This method is applicable for deposition of films of polymeric materials. In all these cases, film formation takes place by evaporation of the solvent from a polymer solution. In dip coating, the substrate is dipped into the polymer solution, taken out, and dried by allowing the solvent to evaporate, leaving behind a solid polymer film on the substrate. The thickness of the coating depends on the viscosity of the solution, the rate of solvent evaporation, and the angle and rate at which the substrate is taken out. Films of thickness as small as 50–100 Å may be deposited by using very dilute and low-viscosity solutions. Films may also be formed by spinning the solution on a rotating substrate. This gives rise to more uniform films. Very thick (~ 100 μm) self-supporting films can be formed by pouring the solution on a horizontally placed substrate and leaving it for the solvent to evaporate. The dried film can be stripped off the substrate very easily, in the manner an adhesive tape is removed.

1.6. *Monitoring and Analytical Techniques*

1.6.1. *General Remarks*

As already emphasized, the structure and hence the properties of thin films are dominated by the nucleation and growth processes. These, in turn, are controlled by a host of deposition parameters such as the incident rate, velocity, angular distribution and nature of species of the impinging vapor, composition of the ambient through which the vapor species travel to condense on a substrate, and the substrate temperature. It is widely recognized that the reproducibility of thin film properties is greatly dependent on how precisely these deposition parameters are monitored and controlled. Of course, it is not possible to monitor all the parameters quantitatively (as in the case of sputtering processes), and therefore emphasis is placed on the similarity of conditions prevailing during deposition.

The physical properties of thin films in the case of vapor deposition techniques can be monitored and measured *in situ*. The specialized techniques depend on the type of thin film material and the information sought. The measurement of such properties as resistivity, Hall coefficient, thermopower, dielectric constant, reflectance, transmittance, magnetization, and stresses is described in standard references mentioned at the beginning of this chapter. By depositing films inside a chamber having analytical facilities (Fig. 1.11), the structure and the composition of the film can be monitored.

1.6.2. *Deposition Rate and Thickness Measurement*

Among the most significant thin film parameters in all deposition techniques are the rate of deposition and thickness. Clearly, *in situ* methods, which measure rate of deposition, also measure film thickness by integrating with respect to time. The *ex situ* methods allow the measurement of film thickness only. In principle, measurement of any physical quantity dependent on film thickness can be used for determining the thickness. Generally, however, this is not practical since the thickness dependence of most physical properties of thin films is strongly affected by the microstructure and hence the deposition parameters. The commonly employed methods are classified into four categories.

1.6.2.1. *Mechanical Method*

The vertical movements of a diamond stylus with a tip diameter of a few microns are amplified to measure step heights and surface irregularities of about 20 Å. A precision of ±2% is obtained in the commercially available instruments such as Talystep/Talysurf and Dektak. By using a small (~0.1 g) tip mass, thin films of even soft materials show little or no deformation under the tip.

1.6.2.2. Ionization Method

Based on measurement of the ion current produced by ionization of the vapor species traversing a nude triode-type ionization gauge, this method measures the rate of evaporation or incidence of the vapors. One must assume a value of the condensation coefficient to obtain the rate of deposition and hence the film thickness (by integrating the signal with respect to time). This method allows measurement of deposition rates down to ~1 Å/sec with several percent accuracy and is well suited for automatic monitoring as well as controlling the deposition process.

1.6.2.3. Microbalance (Gravimetric) Method

By determining the mass of a film of well-defined geometry using a microbalance under *in situ* or *ex situ* conditions, the "gravimetric" film thickness is obtained, provided a value for the film density is assumed. A commercially available microbalance, or a quartz-fiber torsion balance, may be used. A sensitivity of $~10^{-8}$ g (corresponding to a fraction of a monolayer of a typical deposit) is attained in well-designed torsion balances.

The most frequently used microbalance is the oscillating quartz crystal. An AT-cut quartz crystal of thickness t has a fundamental resonance frequency

$$f = 1670/t \text{ (mm) kHz}$$

With the deposition of a material on the crystal, its mass and thickness change by dm and dt, so that the frequency changes by

$$df = -(f^2/1670)\, dt = -\frac{f^2}{1670\,\rho \cdot A}\frac{dm}{}$$

where ρ is the density of the deposit and A is the surface area. Keeping in mind both the frequency dependence of the sensitivity and the convenience of physical handling, 5–6-MHz quartz crystals are commonly used. With a sensitivity $~10^{-9}$ g/cm^2, limited by the temperature coefficient of frequency of the same magnitude, the crystal monitor can measure deposition rates as low as $~10^{-2}$ Å/sec. The upper limit of measurable film thickness is determined by the limit of the linear variation of the frequency with additional mass. It is, however, possible to renew the crystal by removing the deposited film.

1.6.2.4. Optical Methods

The optical methods include optical monitoring and the interferometric and polarimetric methods.

1.6.2.4a. *Optical Monitor.* The simplest optical method is that of measuring the optical absorption which depends exponentially on film thickness and the absorption coefficient. It is obviously necessary to know the absorption coefficient in order to be able to monitor the film thickness *in situ.* If a transparent or a slightly absorbing film is deposited on a nonabsorbing substrate of a different refractive index, the optical reflectance and transmittance exhibit oscillatory behavior with thickness for a given wavelength (or vice versa) due to interference between the light reflected from the front and rear surfaces of the film. The film thickness can thus be monitored *in situ* by monitoring the reflectance and/or transmittance of the film during deposition.

The film thickness can be determined *ex situ* from the maxima and minima of reflectance or transmittance, which occur at intervals given by

$$2nt = m\lambda$$

where n_f is the refractive index of the film of thickness t, λ is the wavelength of light, and m is the order of the maximum and the minimum. Note that all interference techniques measure "optical thickness" (nt), so that n, the refractive index, must be known in order to determine t.

By using a spectrophotometer, the interference maxima and minima can be studied as a function of wavelength. If the mth-order maximum occurs at a wavelength λ_1 and the $(m + 1)$th-order at λ_2, we have, for normal incidence,

$$2nt = m\lambda_1 = (m + 1)\lambda_2$$

Therefore,

$$2nt = \lambda_1\lambda_2/(\lambda_1 - \lambda_2)$$

We have assumed n to be the same at both wavelengths. This simple method allows a quick determination of the optical and physical thickness.

1.6.2.4b. *Interferometric Method.* A direct determination of the thickness and surface topography of a film with considerable precision (down to a few angstroms under optimum conditions) is possible by measuring interference fringes produced by reflections from two reflecting surfaces brought into close proximity in an arrangement called an interferometer. One may use Fizeau's two-beam method in which a parallel monochromatic beam of light is split into two perpendicular beams by means of a beam splitter and the interference is obtained between the beams reflected from the film surface and that from a reference flat. In a multiple-beam interferometric arrangement, the Fizeau fringes of equal thickness are obtained in a setup consisting of two slightly inclined optical flats, one of them supporting the film which forms a step on the substrate. When the second optical flat is brought in contact with the film surface and the setup is illuminated with a parallel monochromatic beam at normal incidence, dark fringes corresponding to points of equal air-gap thickness and separated by $\lambda/2$ are seen. The film thickness step produces a step in these fringes. Multiple beam interferometers are also well suited for surface-roughness studies. If, instead of the air wedge, two parallel plates illuminated with white light are used, colored fringes (called fringes of equal chromatic order) will be observed. A displacement in the fringes due to the film step is utilized to measure thickness.

It should be pointed out that one may also use a sheet of monochromatic X rays at glancing angle to obtain interference fringes between the beams reflected at the air–film and the film–substrate interfaces. This technique has been used for measuring thickness of metal films of thickness as low as 250 Å.

1.6.2.4c. *Polarimetric Method* (*Ellipsometry*). A plane-polarized light reflected from an absorbing substrate assumes elliptical polarization (except when incident normally or at grazing angle, where it is linearly polarized). The measurement of the ellipticity of the reflected beam allows the determination of surface film thickness and its optical constants. Sophisticated commercial instruments, called ellipsometers, are now available. These make it possible to detect monolayers of deposits.

1.6.2.5. *Radiation Methods*

Besides photons, one may use α- or β-rays to measure film thickness by absorption studies. Back-scattering of these radiations (obtained from appropriate radioactive isotopes) also provides a simple and nondestructive technique.

Secondary X rays (x-ray fluorescence) in a thin film may be excited by using a white x-ray radiation or an energetic electron-beam source (as in the scanning electron microscope and the electron microprobe analyzer). The intensity of the excited characteristic radiations is a measure of the number of corresponding atoms. Clearly, it is a good technique for measuring small film thicknesses of multilayer or alloy films.

The important characteristics of the major techniques are summarized in Table 1.2.

1.6.3. *Structural Analysis*

Structural analysis involves the determination of the crystal structure, size and distribution of the grains, topographic details, and structural defects. Because of the strong interaction of electrons with atoms, the high electron-diffraction intensities provide the most sensitive and useful tool for the determination of the crystal structure of films down to monolayer thicknesses. High-energy transmission and reflection electron diffraction techniques are fairly standard tools in any materials science laboratory today. Low-energy electron diffraction (LEED) is a specialized tool in which the diffraction process is characteristic of the few monolayers at the surface and thus provides information on the two-dimensional symmetry and atomic arrangements on the surface of a material. It should, however, be pointed out that quantitative interpretation of the LEED patterns continues to be a challenging job.

As compared with ~1% accuracy of determination of lattice constants with electron diffraction, significantly better accuracies are possible with x-ray diffraction, due to larger diffraction angles. Due to smaller x-ray diffraction intensities, it is necessary to use thicker (>1 μm) films in this case. With the advent of high-intensity x-ray sources such as the rotating anode and synchrotron, x-ray analysis of thin films is becoming competitive with electron diffraction.

High-resolution transmission electron microscopy, in dark and bright field modes, provides the most indispensable technique for the study of microstructure and defect structure of thin films. The resolution of images obtained depends much on the thin film material, thickness, microstructure, electron beam energy, and the electron optics used. Under optimum conditions, using single-crystal specimens, line resolutions ~3 Å and lattice images are easily obtained. By examining a thin epitaxial film deposited on a thin single-crystal substrate of suitable lattice match, moire patterns are formed. The study of these patterns during *in situ* deposition of a film provides the most direct and vivid demonstration of the creation and annihilation of a host of microscopic structural defects.

Surface topography is best studied by electron microscopy of the replica of the thin film surface, by scanning electron microscopy, and by x-ray topographic techniques. Depending on the replication technique and the materials used, image resolutions ~20 Å are attainable. The scanning electron microscope (SEM) is useful only for examining rough surfaces. High-resolution scanning-transmission combination microscopes (STEM) with a resolution ~20 Å in the scanning mode are now available. The x-ray

Table 1.2. Some Common Methods for Deposition Rate, Film Thickness, and Thickness Profile Measurements (after Ref. 8)

Method	Materials	Maximum sensitivity	Maximum thickness	Control automation	Remarks
Resistance	Metals	~1%	~1 μm	Yes	Convenient; empirical thickness–resistance relation required
Ionization gauge	All	<1 Å/sec	None	Yes	Good rate monitor; compensation for residual gas pressure required
Microbalance (gravimetric)	All	~1 Å/cm²	None	Possible	Simple, rate and thickness monitor
Quartz-crystal oscillator	All	≪1 Å/cm²	~1 μm (depends on density)	Yes	Most useful, simple, rate, thickness and vapor distribution monitor
Stylus	All	~20 Å	None	No	Rapid, absolute, thickness and thickness profile
X-ray emission	All (elements)	~100 Å	Depends on excitation agent	No	Simple, relative thickness and thickness profile, electron-probe microanalyzer well suited

Method	Films	Sensitivity	Thickness range		Remarks
Absorption of X, α, and β rays	All	Depends on absorption coefficient	~100 μm	No	Simple, rapid scan, thickness profile
Optical density	All	1%	Transparent films	Possible	Relative, rapid, continuous scan, thickness profile
Photometric	Dielectrics	λ/300	Many μm	Yes	Indispensable for multilayer λ/4 dielectric films
Color of films	Dielectrics	~100 Å	~1 μm	No	Subjective method
Spectrophotometric	Dielectrics	λ/100	Many μm	No	Useful for thick semiconductor and dielectric films
Interferometric	All	~2 Å	Several μm	No	Most accurate, absolute, highly reflecting surface or overcoat required in a two-beam method where sensitivity is limited to ~50 Å
Polarometric (ellipsometric)	Dielectrics (primarily)	<1 Å		No	Tedious, extensive calculations

topographic method has been used to obtain sharp (~100 Å resolution) images of dislocations in thick epitaxial films.

1.6.4. Composition Analysis

With the increasing interest in the use of multilayered and multicomponent material films for microdevices, microscopic analysis of composition, its depth profile and spatial distribution (called *chemical mapping*), and the chemical state of the atoms along with its depth profile are assuming enormous importance in the device technology area today. Conventional chemical analytical techniques such as gravimetric, volumetric, polarographic, chromatographic, colorimetric, spectrophotometric, and optical spectrography (arc, plasma, or flame excited) continue to be utilized for microspecimens. However, in most cases, milligram specimens are required. The detection limit depends on the technique and ranges from a fraction of a percent in volumetric methods to $\sim 10^{-4}\%$ in spectrographic techniques.

The requirements of a detailed microanalysis with a high depth and spatial resolution have led to the development of a host of surface analytical techniques. These techniques are based on the analysis of the emitted characteristic photons, electrons (Auger and secondary), ions, or atoms/molecules from a surface excited by an appropriate "probe" radiation (for example, photons, electrons, or ions). Various combinations of the probe and the emitted radiations give rise to a number of techniques. Of these, the ones which are of direct interest for thin film analysis are the electron microprobe analyzer (EMA), scanning Auger microprobe or Auger electron spectroscopy (SAM or AES), electron spectroscopy for chemical analysis (ESCA), the secondary ion mass spectrometer (SIMS), and Rutherford back-scattering (RBS). In the EMA technique, characteristic x-rays of the elements in the film are excited by an impinging electron beam and are analyzed for composition analysis. In the AES technique, a primary electron beam knocks off an electron from a core level of an atom following which there is an internal rearrangement of the electrons, as a result of which a second electron, called the Auger electron, is emitted. The measurement of the number and energy of these characteristic electrons emerging from within the escape depth (~30 Å) on the surface allows identification of the element and measurement of its concentration. The ESCA technique employs a soft x-ray source (for example, Al K_α, Mg K_α) to remove electrons from various energy shells of well-defined binding energies. The measurement of the energy of these electrons provides a direct value of the binding energies of various core levels and information on the position of the valence band and its density of states. The measurement of the chemical shift—the shift of the core-level binding energy of an element in its compound–alloy form as compared with that in the elemental form—helps to establish the chemical state of the atom. The SIMS technique uses the principle of analyzing the charge-to-mass ratio of the ions knocked off from the specimen surface due to bombardment by an energetic beam of heavy ions (e.g., argon). Combined with simultaneous ion beam etching, the SAM, ESCA, and SIMS techniques enable depth profiling of the specimens.

When a surface is bombarded with H^+ or He^+ ions in the energy range 100 keV to 5 MeV, a small fraction ($\sim 10^{-6}$) of the incoming particles undergoes Rutherford collisions and is back-scattered. Energy analysis of the elastically scattered ions yields not only the mass of the scattering atom (and hence its identification) but also the depth from the surface from where the collision has taken place. By varying the energy of the incident particles, RBS gives elemental atomic ratios and the depth distribution of elements without standards and without destroying the samples. Due to larger scattering cross section for increasing atomic number, this technique is well suited for trace analysis of heavy elements.

Each analytical technique has its own sensitivity and resolution limits and provides characteristic information. The prominent features of these techniques are summarized in Table 1.3. It is clear that no single technique provides uniquely complete information. It is therefore essential to employ two or more techniques (as shown in Fig. 1.11) for sequential analysis of the same specimen. Since each analytical facility is sophisticated and expensive, it is important to have a combination best suited to the particular needs of an R/D problem. Which combination to employ for a particular problem can be decided only by a detailed analysis of the specifications of the analytical techniques and the required information.

1.7. Microfabrication Techniques

For the thin film to be useful, it is necessary to deposit or fabricate[36–40] it in a particular geometrical shape of precise dimensions. Thin film interconnections in integrated circuits and thin film integrated optical circuits represent some of the demanding applications of microfabricated and structured thin films of different materials that perform a variety of functions.

The simplest technique to structure films is to deposit films through a patterned mask. The resolution of the thin film pattern is determined by the definition of the mask and its proximity to the substrate. Thin metal-foil masks obtained by photolithographic techniques can yield a line resolution $\sim 10 \ \mu$m. For better resolution and smaller line widths, it is necessary to use lithographic techniques to produce an etched pattern.

Table 1.3. Comparison of Various Techniques for Compositional Analysis

Method	Probe	Detected species	Probe diameter	Depth resolution	Detection limit	Quantitative analysis	Chemical state	Imaging capability	Depth profiling	Specimen destruction
AES	e-beam	Auger electrons	0.2 μm	5.40 Å	0.1 at.%	Yes	Yes	Good	Yes	No
ESCA	X rays	Core and valence electrons	5 mm	10–100 Å	0.5 at.%	Yes	Good	Difficult	Yes	No
SIMS	Ions	Ions	10 μm	<10 Å	1 ppm	Very difficult	Yes	Yes	Yes	Yes
X-ray fluorescence	X rays or γ rays	X rays	1 cm	2–3 μm	0.01 at.%	Yes	No	No	No	No
EMA	e-beam	Characteristic X rays	≤100 Å	1–2 μm	0.1 at.%	Yes	No	Yes	Yes	No
RBS	α-particles	α-particles	1 mm	100 Å	10^{-3} at.%	Yes	No	No	Yes	No

Photolithography is a replication technique. Here, a master mask is created in a photographic emulsion film, or a thin film (e.g., of chromium or iron oxide) coated glass or quartz substrate. The film to be patterned is first coated with a thin ($\leqslant 1$ μm) layer of a photoresist material by a standard liquid spinning technique. It is then illuminated with a collimated parallel beam of light from a UV source, through the aligned master mask which is either pressed against the film (contact printing) or is in close proximity (projection printing). Depending on the nature of the photoresist used, it hardens (negative photoresist) or softens (positive photoresist) on being exposed to the UV radiation through the transparent regions of the mask. On postbaking, the hard regions are fixed to become resistant to chemical etching in an appropriate solvent, called a *developer*. Thus, depending on the nature of the photoresist used a positive or negative image of the master mask can be created in the required film, as shown in Fig. 1.15. The exposed parts may then be etched away to the desired depth by wet chemical, dry plasma, or ion-beam techniques. The remaining photoresist on the film is subsequently removed with the help of a suitable chemical solution.

The resolution of a UV lithographic process depends on (1) the resolution of the mask, (2) the resolution of the photoresist, (3) the optical abberations and depth of focus in the exposure field, (4) the penumbral effect due to the physical separation of the mask and the film, and (5) the diffraction effects. Submicron-resolution masks can now be made with x-ray and e-beam lithographic techniques. A large variety of photoresists with resolution in micron and submicron dimensions have been developed and are commercially available. Optical abberations and penumbral effects can

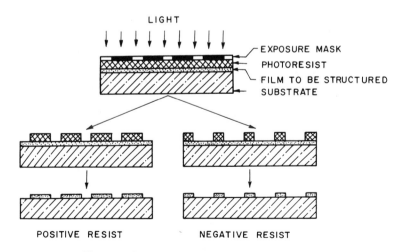

Figure 1.15. Pattern generation by photolithography.

be minimized. Finally, however, the line resolution is limited by diffraction and is thus given by the Airy disk diameter

$$D = 2\lambda F$$

where λ is the wavelength of radiation used and F is the numerical aperture. The depth of focus is

$$d = 2 \times 2.44\lambda F^2$$

By optimizing the resolution and depth of focus requirements, optical lithography with a resolution ~ 1 μm is possible and is being realized in laboratory practice.

For submicron resolution, it is clear that we must decrease the radiation wavelength. This has led to very exciting research and development in x-ray, electron-beam, and ion-beam lithographies. Resist materials for these lithographies are currently under extensive investigations. Among the promising new materials are amorphous chalcogenide films (see Section 3.7.3 of Chapter 3). Based on the radiation-induced contraction effect discovered in our laboratory, we have demonstrated submicron-resolution reprographic patterns in obliquely deposited films of amorphous Ge–Se films (see Section 3.7.3 of Chapter 3) using photons, electrons, and ion beam. In all the cases, submicron-resolution masks must necessarily be prepared by electron-beam lithography, using a raster or vector scanning electron beam of nanometer dimensions. With the present technology, x-ray and ion-beam lithographies are possible only by proximity printing through masks. In contrast, the e-beam technique allows direct writing. Direct writing by ion beams in the micron range of resolution has also been demonstrated. In these lithographies, the ultimate limit attainable in ultrathin noncrystalline films of resist materials is set by the scattering and secondary excitation processes due to the incident radiation. For example, one should be able to attain a resolution of ~ 200 Å in suitable chalcogenide films. Based on the rapid developments in the filed of microlithography, it is concluded that submicron film structures will be widely available in this decade.

When all the three dimensions of a thin film are in the submicron range, a wide variety of new physical phenomena are expected to occur. These include mean-free-path and quantum-size effects, electromigration, hot electron transport, enormous surface reactivity, unusual diffusion and alloying effects, metastable structures, etc. These exciting areas of solid state physics, chemistry, and metallurgy are already emerging into a new frontier, called *microscience*.

2

Thin Films in Optics

Before we go into the description of various optical coatings, we would like to familiarize the reader with the basic optics of thin films that is necessary for the understanding and design of these coatings.

2.1. Optics of Thin Films

When electromagnetic radiation is incident on a boundary separating two media of optical admittances N_0 and N_1, part of the incident radiation is reflected back into the incident medium. The energy reflection coefficient or the reflectance at the interface for an angle of incidence θ is given by

$$R = \left(\frac{\eta_0 - \eta_1}{\eta_0 + \eta_1}\right)^2 \qquad (2.1)$$

where η represents the effective optical admittance and is given by

$$\eta = \begin{cases} N \cos \theta & \text{for TE waves (or } s \text{ waves)} \\ N/\cos \theta & \text{for TM waves (or } p \text{ waves)} \end{cases}$$

N, the optical admittance, often called the complex refractive index, is given in terms of the refractive index n and the extinction coefficient k as

$$N = n - ik$$

The most general system of thin films consists of an assembly of l layers ($j = 1$ to l, with $j = 1$ as the outermost) with optical admittance N_j and thickness t_j on a substrate of admittance N_s. If Y is the equivalent optical admittance of this sytem, then the normal reflectance in a medium of admittance N_0 is given by

$$R = \left(\frac{N_0 - Y}{N_0 + Y}\right)^2 \tag{2.2}$$

An elegant way of calculating Y for this assembly of thin films is the matrix method developed by Herpin[1] and Weinstein.[2] According to this method

$$Y = C/B \tag{2.3}$$

where B and C are the elements of a 1×1 matrix, called the characteristic matrix of the assembly and defined as

$$\begin{bmatrix} B \\ C \end{bmatrix} = \left\{ \prod_{j=1}^{l} \begin{bmatrix} \cos \delta_j & i \sin \delta_j/\eta_j \\ i\eta_j \sin \delta_j & \cos \delta_j \end{bmatrix} \right\} \begin{bmatrix} 1 \\ \eta_s \end{bmatrix} \tag{2.4}$$

where

$$\delta_j = (2\pi/\lambda)(N_j t_j \cos \theta_j) \tag{2.5}$$

is the effective phase thickness of the jth layer; λ is the wavelength of the incident radiation in vacuo, and θ_j is the angle of refraction in the jth layer and is related to the angle of incidence θ by Snell's law:

$$N_0 \sin \theta = N_j \sin \theta_j \tag{2.6}$$

$N_j t_j$ and $N_j t_j \cos \theta_j$ are called the optical thickness and the effective optical thickness, respectively. Equations (2.2)–(2.4) are of utmost importance and form the basis of almost all thin film calculations. The calculation of R using the matrix method is illustrated below for the case of a single, nonabsorbing film of refractive index n_1 and thickness t_1 on a substrate of index n_s. The characteristic matrix of the film–substrate combination is

$$\begin{bmatrix} \cos \delta_1 & i \sin \delta_1/\eta_1 \\ i\eta_1 \sin \delta_1 & \cos \delta_1 \end{bmatrix} \begin{bmatrix} 1 \\ \eta_s \end{bmatrix} = \begin{bmatrix} \cos \delta_1 + (i\eta_s/\eta_1) \sin \delta_1 \\ \eta_s \cos \delta_1 + i\eta_1 \sin \delta_1 \end{bmatrix}$$

Therefore

$$R = \frac{(\eta_0 - \eta_s)^2 \cos^2 \delta_1 + [(\eta_0\eta_s/\eta_1) - \eta_1]^2 \sin^2 \delta_1}{(\eta_0 + \eta_s)^2 \cos^2 \delta_1 + [(\eta_0\eta_s/\eta_1) + \eta_1]^2 \sin^2 \delta_1} \tag{2.7}$$

If λ_0 is the wavelength at which the layer is a quarter wavelength in optical thickness, that is, $n_1 t_1 = \lambda_0/4$, then, for normal incidence, $\delta_1 = \pi/2$ at $\lambda = \lambda_0$, and therefore the normal reflectance is given by

$$R = \left(\frac{n_0 n_s - n_1^2}{n_0 n_s + n_1^2}\right)^2 \quad \text{for } n_1 t_1 = (2m + 1)\frac{\lambda_0}{4} \tag{2.8a}$$

$$= \begin{cases} R_{\min} & \text{if } n_s > n_1 > n_0 \tag{2.8b} \\ R_{\max} & \text{if } n_s < n_1 > n_0 \tag{2.8c} \end{cases}$$

If the layer is a half wave in optical thickness, then at normal incidence $\delta_1 = \pi$, and R reduces to

$$R = \left(\frac{n_0 - n_s}{n_0 + n_s}\right)^2 \quad \text{for } n_1 t_1 = (2m + 2)\frac{\lambda_0}{4} \tag{2.9a}$$

$$= \begin{cases} R_{\max} & \text{if } n_s > n_1 > n_0 \tag{2.9b} \\ R_{\min} & \text{if } n_s < n_1 > n_0 \tag{2.9c} \end{cases}$$

= reflectance of the bare substrate.

Equation (2.7) shows that the normal reflectance exhibits an oscillatory behavior with optical thickness in wavelengths, as shown in Fig. 2.1. A

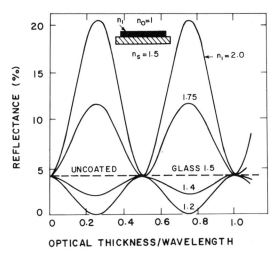

Figure 2.1. Theoretical variation of normal reflectance R (at air side) with optical thickness ($n_1 t_1$) in wavelengths, for quarter-wave-thick (at λ_0) films of different refractive indices (n_1) on a glass substrate of index 1.5 (after Ref. 45).

very interesting result which follows from the above two sets of equations is that a $\lambda/4$ coating of a material on a substrate diminishes or enhances its reflectance to a value given by Eq. (2.8a), depending on whether $n_1 < n_s$ or $n_1 > n_s$. This principle is utilized for the construction of antireflection or reflection coatings described in the following sections. It is also clear from Eqs. (2.9a)–(2.9c) that the normal reflectance of a substrate remains unaltered by coating a half-wave layer of any material on it. For this reason, a $\lambda/2$ layer is sometimes called an "absentee" layer.

2.2. Antireflection Coatings (AR Coatings)

The normal reflectance at an interface between two media, as given by Eq. (2.1), corresponds to 4% for an air–glass ($n = 1.5$) and 36% for an air–Ge ($n = 4.0$) interface in the region of transparency. The light so reflected can greatly hamper the performance of an optical system both on the grounds that it is lost from the main beam, thus weakening the image, and also because some part of it reaches the image plane appearing as a veiling glare or ghost, thus reducing the contrast. Reduction of the reflectance and enhancement of the transmittance may be achieved by the use of suitable surface coatings called *antireflection coatings*. This antireflection property is also called blooming. Because of their wide range of applications, AR coatings have been the subject of extensive investigations. For a review of the literature and systematic discussion of AR coatings, the reader is referred to excellent review articles by Macleod,[3] Cox and Hass,[4] and Mussett and Thelen.[5]

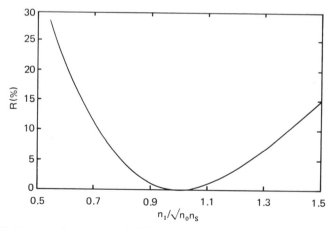

Figure 2.2. Normal reflectance (at air side) of a quarter-wave-thick film of refractive index n_1 on a substrate of index n_s, plotted as a function of $n_1/(n_0 n_s)^{1/2}$.

2.2.1. *Single-Layer AR Coatings*

We have seen that a $\lambda/4$ coating on a substrate of higher index minimizes its normal reflectance to a value given by Eq. (2.8a), because of destructive interference between the rays reflected from the top and the bottom surfaces of the film. Clearly, the condition for the minimum reflectance to be zero is

$$n_1^2 = n_0 n_s$$

A universal curve for R vs. $n_1/(n_0 n_s)^{1/2}$ is shown in Fig. 2.2. It shows that for values of R other than zero, there exist two values of the film index, $n_1' < n_s$ and $n_1'' > n_s$, such that $n_1' n_1'' = n_0 n_s$, which give the same reflectance with identical spectral dependence. Equations (2.5) and (2.7) show that for nonnormal incidence, the minimum reflectance increases and shifts to shorter wavelengths because of a decrease in the effective phase thickness of the layer, as illustrated in Fig. 2.3 for MgF_2 AR coating on glass. For high-index bloomed surfaces, the reflectance shows a slightly stronger dependence on angle of incidence and rises much more steeply around the minimum than for low-index materials.

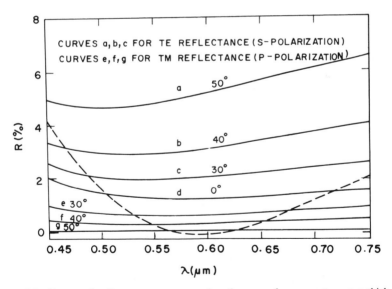

Figure 2.3. Computed reflectance versus wavelength curves for a quarter-wave-thick (at normal incidence and $0.6\ \mu m$) AR coating of MgF_2 on glass ($n_1 = 1.38$, $n_s = 1.52$) at various angles of incidence. Also for comparison is shown (dotted) the reflectance curve for a quarter–quarter double-layer AR coating [$n_1 = 1.38$ (MgF_2), $n_2 = 1.70$ (SiO), and $n_s = 1.52$ (glass)] (after Ref. 3).

The effectiveness of a single-layer AR coating is limited by the availability of materials, and, in practice, perfect blooming cannot be realized for substrates with index lower than 1.9. Nevertheless, a worthwhile suppression of R, sufficient for many purposes for substrates of index as low as 1.5, can be obtained even with the available materials. For example, a material with $n = 1.23$, for perfect blooming of crown glass ($n = 1.52$) is not available, yet normal reflectance can be reduced to 1.33% with MgF_2 ($n = 1.38$) and to 0.75% with cryolite ($n = 1.34$) compared to 4.1% for uncoated glass surface. It should be noted that even with the right material, zero reflectance is obtained only at a wavelength λ_0 at which the layer is a quarter wave in optical thickness. More effective blooming can be obtained using two or more layers.

2.2.2. Double-Layer AR Coatings

The conditions for zero reflectance for a double layer can be very easily determined using the matrix method. The optical admittance of a double layer is

$$Y = \frac{i\eta_1 \sin \delta_1(\cos \delta_2 + (i\eta_s/\eta_2) \sin \delta_2) + \cos \delta_1(\eta_s \cos \delta_2 + i\eta_2 \sin \delta_2)}{\cos \delta_1[\cos \delta_2 + (i\eta_s/\eta_2) \sin \delta_2] + (i \sin \delta_1/\eta_1)(\eta_s \cos \delta_2 + i\eta_2 \sin \delta_2)} \quad (2.10)$$

Applying the condition for zero reflectance, that is, $Y = \eta_0$, and equating the real and imaginary parts, we obtain

$$\tan \delta_1 \tan \delta_2 = \frac{\eta_1 \eta_2(\eta_s - \eta_0)}{\eta_1^2 \eta_s - \eta_0 \eta_2^2} \quad (2.11)$$

and

$$(\tan \delta_2/\tan \delta_1) = \frac{\eta_2(\eta_0 \eta_s - \eta_1^2)}{\eta_1(\eta_2^2 - \eta_0 \eta_s)} \quad (2.12)$$

Thus,

$$\tan^2 \delta_1 = \frac{(\eta_s - \eta_0)(\eta_2^2 - \eta_0 \eta_s)\eta_1^2}{(\eta_1^2 \eta_s - \eta_0 \eta_2^2)(\eta_0 \eta_s - \eta_1^2)} \quad (2.13)$$

and

$$\tan^2 \delta_2 = \frac{(\eta_s - \eta_0)(\eta_0 \eta_s - \eta_1^2)\eta_2^2}{(\eta_1^2 \eta_s - \eta_0 \eta_2^2)(\eta_2^2 - \eta_0 \eta_s)} \quad (2.14)$$

The correct pair of δ_1, δ_2 determined from (2.13) and (2.14) must also satisfy either of the two preceding equations. At the wavelength λ_0 where both layers are quarter wave thick and at normal incidence, $\delta_1 = \delta_2 = \pi/2$, and therefore from Eq. (2.10),

$$Y = n_1^2 n_s / n_2^2 \tag{2.15}$$

giving,

$$R = \frac{n_0 n_2^2 - n_1^2 n_s}{n_0 n_2^2 + n_1^2 n_s} \tag{2.16}$$

Two approaches can be used for designing double-layer AR coatings. In the first, use is made of the available materials with prefixed values of the indices, and the thicknesses required to give zero reflectance are computed using Eqs. (2.13) and (2.14), as a result of which the thicknesses obtained are awkward fractions of $\lambda/4$. Another approach is to use layers with optical thicknesses that are an integral multiple of $\lambda/4$ and compute the refractive indices required to give zero reflectance.

2.2.2.1. Case 1: Each Layer Thickness $\lambda/4$ (Quarter–Quarter Coating)

In this case, for normal incidence (i.e., $\delta_1 = \delta_2 = \pi/2$), the condition for zero reflectance as obtained from Eqs. (2.13) and (2.14) is

$$n_1^2 n_s / n_2^2 = n_0 \tag{2.17}$$

For crown glass ($n = 1.52$) as the substrate and air ($n = 1.00$) as the incident medium, if the first layer is MgF_2 ($n = 1.38$), the index of the second (inner) layer must be $n_2 = (n_1^2 n_s / n_0)^{1/2} = 1.70$.

The normal reflectance of this type of coating, first studied by Muchmore,[6] is shown by the dotted curve in Fig. 2.3, obtained using SiO ($n = 1.7$) as the second material. Yadava and Chopra[7] have shown that a $50:50$ mixture of ZnS and MgF_2, which has a refractive index of 1.7, may also be used in place of SiO to give a similar reflectance curve. Thus we see that a double-layer AR coating makes it possible to obtain zero reflectance on glass with durable materials. MgF_2 becomes more effective now because the substrate appears to have a higher index. However, although the reflectance can be made zero at the design wavelength, the reflectance of the double layer rises much more steeply near the minimum than that of a single layer, because the inside layer next to the substrate has an index higher than the substrate. For materials having a high refractive

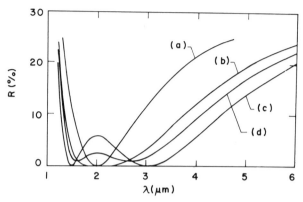

Figure 2.4. Reflectance-versus-wavelength curves for various types of AR coatings (quarter-wave-thick at 2 μm) on Ge (n_s = 4.0): (a) single layer (n_1 = 2.0); (b) double layer (n_1 = 1.414, n_2 = 2.829); (c) double layer (n_1 = 1.6, n_2 = 2.5); (d) double layer (n_1 = 1.35, n_2 = 2.3) (after Ref. 4).

index, such as infrared transparent materials, Eq. (2.17) for zero reflectance yields the so-called "stepdown design", i.e., $n_0 < n_1 < n_2 < n_s$. Here since no layer has an index higher than the substrate, zero reflectance is obtained over a region much broader than that of a single-layer coating, as shown in Fig. 2.4 (curves a and b).

2.2.2.2. *Case 2: Equal Optical Thickness, Nonquarter Coatings*

A much broader region of low reflectance can be obtained with this kind of design. Here $\delta_1 = \delta_2 = \delta$. Putting this in Eq. (2.12) we obtain the condition for zero normal reflectance as

$$n_0/n_1 = n_2/n_s \qquad (2.18)$$

Substituting Eq. (2.18) in Eq. (2.11), we obtain the equation for the thickness of the two layers as

$$\tan^2 \delta = \frac{n_1^2 n_2 - n_1 n_0^2}{n_1^3 - n_0^2 n_2} \qquad (2.19)$$

Equation (2.19) has two roots, showing that zero reflectance is obtained at two wavelengths given by these two roots. Also, at the wavelength where both layers are quarter wave thick, a maximum in reflectance is obtained, with the value being given by Eq. (2.16). The reflectance of this type of coating showing the double zero characteristic is illustrated in Fig. 2.4 (curve c). Further, a compromise coating may give even a more extended range of blooming than with ideal index values, as shown by curve d in Fig. 2.4. The increase in reflectance with increasing angle of incidence using this design is relatively small. Unfortunately, this design cannot be used in the visible region because of the nonavailability of suitable low-index materials.

2.2.2.3. *Case 3: Inner Layer* $\lambda/2$, *Outer Layer* $\lambda/4$

The condition of zero reflectance for the general case where the inner layer is twice as thick as the outer layer as derived from Eqs. (2.11)–(2.14) is given by

$$n_2^3 - \frac{1}{2}\frac{n_2 n_s}{n_0 n_1}(n_0^2 + n_1^2)(n_1 + n_2) + n_1 n_s^2 = 0 \qquad (2.20)$$

Also, the phase thickness δ_1 satisfies the relation

$$\cos^2 \delta_1 = \frac{1}{2}\left[1 - \frac{(n_0 + n_1)(n_1 n_s - n_2^2)}{(n_0 - n_1)(n_1 + n_2)(n_2 + n_s)}\right] \qquad (2.21)$$

This type of coating has two zeros at wavelengths determined by the two roots of Eq. (2.21). The simplest case is that where the inner layer is $\lambda/2$ thick and the outer layer is $\lambda/4$ thick at the design wavelength, where a minimum in reflectance, given by Eq. (2.8a) for a single $\lambda/4$ layer, is obtained.

2.2.2.4. *Case 4: Layer Thicknesses Are Nonintegral Multiples of* $\lambda/4$

In this case the values of the indices of both layers are known, and the layer thicknesses required to give zero reflectance are calculated using Eqs. (2.13) and (2.14). The ranges of the indices n_1 and n_2 for which δ_1 and δ_2 are real and zero reflectance can be obtained are beautifully illustrated by Schuster's diagram (Fig. 2.5) for $n_0 = 1$ and $n_s = 4.0$. The shaded areas represent the regions where zero reflectance can be obtained with real values of δ_1 and δ_2. The horizontal and vertical boundaries correspond to single-layer $\lambda/4$ AR coatings, and the diagonal boundary corresponds to $\lambda/4$–$\lambda/4$ AR coatings. The dashed curve corresponds to equal-thickness, nonquarter AR coatings. For high-index materials, the shaded rectangle, where $n_1 \geqslant (n_s)^{1/2}$ and $1 \leqslant n_2 \leqslant (n_s)^{1/2}$, gives very interesting AR coatings in which the total optical thickness of the double layer is less than $\lambda/4$ at the design wavelength and the high-index material forms the outer layer in contrast to all the previous designs, thus protecting the less robust low-index layer.

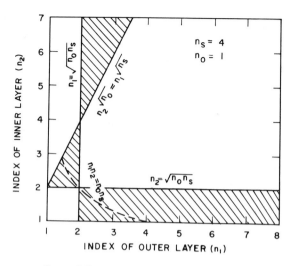

Figure 2.5. Schuster's diagram (after Ref. 4).

2.2.3. Multilayer and Inhomogeneous AR Coatings

A much broader region with lower and more uniform reflectance, as desired for some applications, is obtained by the use of multilayer coatings. We do not intend to discuss the analysis here since it is similar to that for double-layer coatings, already discussed in detail. However, we have included these in Table 2.1, which shows some of the important practical AR coatings for various substrates.

Addition of more and more layers to the multilayer quarter-wave AR coating steadily improves the performance both from the point of view of bandwidth and of maximum reflectance in the low reflectance region. In the limit the number of layers tends to infinity and the multiple layers become indistinguishable and behave as a single inhomogeneous layer with optical admittance smoothly varying from that of the substrate to that of the incident medium. This forms a perfect AR coating for all wavelengths shorter than twice the optical thickness of the film. The performance falls off as the wavelength increases further, and at a wavelength four times the optical thickness the coating ceases to be effective. Because of the non-availability of suitable low-index materials, any inhomogeneous coating must terminate with an index of about 1.35. The performance of an inhomogeneous AR coating (physical thickness 1.2 μm) of Ge–MgF$_2$ on Ge substrate is shown[8] in Fig. 2.6.

Table 2.1. Practical AR Coatings for Some Common Substrates[a]

Substrate	n_s (in the region of transparency)	n_1	n_2	n_3	n_4
Glass	1.51	1.38			
		1.38	1.7		
		1.47	2.14	1.80	
		1.38	1.82	2.20	2.96
Ge	4.1	2.2			
		1.57	3.3		
		1.35	2.2	3.3	
		1.38	1.82	2.2	2.96
Si	3.5	1.85			
		1.35	2.2		
		1.38	1.86	2.56	
ZnS	2.2	1.59			
InAs	3.4	1.85			

[a] All the coatings are quarter-wave thick, i.e., $nt/\lambda = 0.25$ ($n_0 = 1$ for all cases).

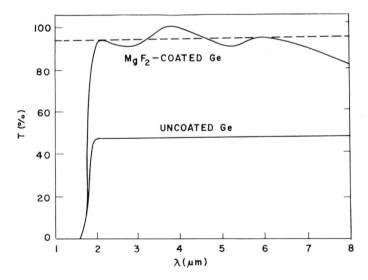

Figure 2.6. Transmittance of a Ge substrate, coated on both sides with inhomogeneous Ge–MgF$_2$ film with physical thickness 1.2 μm (after Ref. 8).

2.3. Reflection Coatings

2.3.1. Metal Reflectors

The use of thin metal films as reflecting surfaces has been known for a long time. The performance of some common metal reflectors is shown[9] in Fig. 2.7. The wide range of constant reflectance of metal films is very useful in applications such as neutral mirrors, neutral beam splitters, and neutral density filters. Metal layers can be protected and their reflectance enhanced over a limited spectral range by using suitable dielectric coatings such as SiO, SiO_2, and Al_2O_3. Whereas a single dielectric layer generally slightly lowers the reflectance, a pair of quarter-wave dielectric layers in which the outer layer is the high-index material, that is, $n_1 > n_2$, boosts the reflectance. Examples of such pairs for Al are MgF_2–CeO_2 and MgF_2–ZnS. The higher the ratio n_1/n_2 and the number of such $\lambda/4$ pairs, the greater is the increase in reflectance.

When metal mirrors are used for partial reflection and transmission, a considerable loss of light takes place by absorption. Although this permits a reasonable level of performance in very simple systems and in two-beam interferometers, it renders them unsuitable for more sophisticated applications. Such highly reflecting systems with very low absorption are possible through the use of dielectric films.

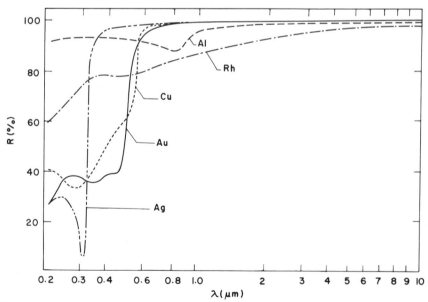

Figure 2.7. Reflectance of freshly deposited films of Al, Cu, Au, Rh, and Ag as a function of wavelength (after Ref. 9).

2.3.2. All-Dielectric Reflectors

The use of single quarter-wave films of index higher than the substrate for boosting its reflectance has already been mentioned in Section 2.2.1. Consider now an assembly of an odd number of quarter-wave-thick dielectric layers of indices $n_1, n_2, \ldots, n_{2p+1}$ on a substrate of index n_s, the $(2p + 1)$th layer being closest to the substrate. The optical admittance of this assembly at the design wavelength λ_0 is given, using the characteristic matrix, by

$$Y = \frac{\eta_1^2 \eta_3^2 \eta_5^2 \cdots \eta_{(2p+1)}^2}{\eta_2^2 \eta_4^2 \eta_6^2 \cdots \eta_{2p}^2 \eta_s} \qquad (2.22)$$

This shows that the optical admittance and hence the reflectance is high if $n_1, n_3, n_5, \ldots$ are all high and $n_2, n_4, n_6, \ldots$ are all low, corresponding to a design of the type: substrate–HLHL$\cdots$HLH or, in short, S–H(LH)^p, where H and L denote quarter-wave thicknesses of the high- and low-index materials, respectively, and S denotes the substrate. If $n_1 = n_3 = n_5 = \cdots = n_{(2p+1)} = n_H$ and $n_2 = n_4 = n_6 = \cdots = n_{2p} = n_L$, we have

$$Y = (\eta_H/\eta_L)^{2p}(\eta_H^2/\eta_s)$$

Hence, the normal reflectance at λ_0 in air is

$$R = \left[\frac{1 - (n_H/n_L)^{2p}(n_H^2/n_s)}{1 + (n_H/n_L)^{2p}(n_H^2/n_s)} \right] \qquad (2.23)$$

A typical characteristic of this type of quarter-wave stack is shown[10] in Fig. 2.8. The reflectance remains high over specific spectral regions centered around $\lambda_0, \lambda_0/3, \lambda_0/5, \ldots$, and outside these it falls rapidly to a low oscillatory value. Addition of more LH pairs to the stack increases the reflectance within the high-reflectance zones and the number of oscillations outside, without affecting the width of the zones, which is equal to $(4/\pi)\sin^{-1}[(n_H - n_L)/(n_H + n_L)]$ and therefore depends only on the ratio n_H/n_L. Apart from its use as a subunit of various optical coatings, the quarter-wave stack is used in Febry–Perot (FP) interferometers, in cavity reflectors for lasers, optical delay lines, etc. Cold mirrors which combine high reflectance of the stack in the visible with high transmittance of the substrate (like Ge) in infrared are another ingenious application of the quarter-wave stack.

The limited width of the high reflectance zone of a $\lambda/4$ LH stack, which may prove unsuitable for some applications, can be widened by staggering the thickness of the successive layers throughout the stack, so

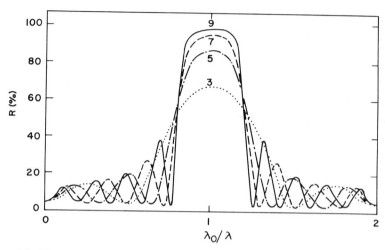

Figure 2.8. Normal reflectance curves of $H(LH)^p$ (p = 3, 5, 7, 9) quarter-wave-thick stack of ZnS (n_H = 2.3) and MgF_2 (n_L = 1.38) on glass (n_s = 1.52) (after Ref. 10).

that the effect is similar to a $\lambda/4$ stack where the design wavelength varies continuously. Another method of widening the high reflectance zone consists of depositing a $\lambda/4$ stack for one wavelength over another for a different wavelength. A low-index layer which is one quarter wave thick at the mean wavelength is inserted between the two stacks to avoid the resulting transmission peak in the high-reflectance zone. A typical design is as follows: $S-H(LH)^p L''(H'L')^p H'$, where $H(LH)^p$ is centered on λ_1, $(H'L')^p H'$ on λ_2, and L'' on $(\lambda_1 + \lambda_2)/2$.

One feature of these broadened band reflectors which makes them unsuitable for wavelength determination in FP interferometers is that the phase of the reflected radiation varies much more rapidly with wavelength than in the case of a simple quarter-wave stack. Another difficulty is the increased sensitivity to absorption losses in the materials, especially in the staggered thickness layers. Nevertheless, broad-band all-dielectric reflectors can be used without difficulty as laser reflectors. Because of these difficulties and because these coatings involve a large number of layers, for high reflectance over wide regions metal layers are still the first choice in those cases where absorption losses can be tolerated.

2.4. Interference Filters

Optical filters based on the phenomenon of interference in thin films are called *interference filters*. According to their characteristics, these can be divided broadly into two categories: (1) edge or cutoff filters and (2) band filters. *Edge filters* are those whose primary characteristic is an abrupt change between a region of transmission and a region of rejection. These can be either short-wave pass or long-wave pass. *Band filters* have a region of transmission bounded on either side by regions of rejection or vice versa, and are accordingly called band-pass filters or band-stop filters. Both can be either of wide-band or narrow-band type. Band-stop or rejection filters are nothing more than the reflection coatings described in the previous section. Unfortunately, very-narrow-band rejection filters are not easily available because the narrower the stop band, the greater is the number of layers required to achieve necessary rejection and the smaller are the thickness tolerances and hence the more serious the reduction in performance due to errors. The construction of various other filters is described in the following sections.

2.4.1. Edge Filters

An ideal edge filter would reject all the radiation below and pass all that above a certain wavelength or vice versa, and is completely characterized by the cut-off wavelength. Real edge filters are, however, not perfect and so alongwith the cut-off wavelength, the slope of the transition region and the extent and average values of the transmission and rejection must also be specified.

The reflectance curve (Fig. 2.8) of a quarter-wave stack $H(LH)^p$ described in the previous section shows that it can be used both as a short-wave-pass and a long-wave-pass filter. The rejection zone of such cut-off filters is not sufficiently wide for some applications where it is required to eliminate all wavelengths shorter or longer than a particular value. The rejection zone of this edge filter can be extended by combining it with an absorption filter. The main defect in the quarter-wave stack is the prominent ripple in the pass region due to mismatching of the optical admittances of the substrate, stack, and medium. This can be considerably reduced by the addition of $\lambda/8$ layers on either side of the stack. It does not take long for one to see that the new design $A/2\ B\ (A\ B)^p\ A/2$ corresponds to a periodic stack of $(p + 1)$ symmetrical multilayers of the form $(A/2\ B\ A/2)$, where A and B can be H or L depending on whether the filter is a short-wave or a long-wave pass. On the basis of the theory developed by Epstein[11] for periodic symmetric multilayers, it has been established that on a glass substrate the assembly $(H/2\ L\ H/2)$ works as a long-wave pass and $(L/2\ H\ L/2)$ as a short-wave pass. The position and width of the high-reflectance zones is the same as in the quarter-wave stack. For obtaining minimum ripple in the pass region using symmetric layer design, the refractive indices n_A and n_B of the coating materials must satisfy the condition[12]

$$n_s n_0 = n_A^3/n_B \qquad \text{for a short-wave pass}$$

and

$$n_s n_0 = n_A n_B \qquad \text{for a long-wave pass}$$

If materials with suitable refractive indices are not available, the ripple can be reduced by adding two AR coatings, one matching the stack to the substrate and the other to the medium. Other less frequent methods of reducing the ripple in the pass region are: (1) adjustment of the thickness of all the layers in the basic period to bring the equivalent admittance of the stack nearer to the optimum value (due to Baumeister and Jenkins[13]), (2) use of inhomogeneous layers on either side of the stack (due to Jacobsson[14]), and (3) use of an equiripple design (due to Young[15]) in which thicknesses are kept constant and the indices varied.

Another difficulty which arises in the case of short-wave pass filters is the limited width of the pass region because of the higher-order reflectance zones present at wavelengths $\lambda_0/3$, $\lambda_0/5$, $\lambda_0/7$ and so on, where the phase thickness of the basic period is $m\pi$, m being an integer. The higher-order reflectance zones can be suppressed if the transmission of the stack at these wavelengths is made equal to unity. This can be achieved by using Thelen's design,[16] in which the basic symmetric period of total optical thickness $\lambda_0/2$ consists of three materials in the form (ABCBA), with the optical thickness of the A and B layers equal. If λ_1 and λ_2 are the wavelengths where suppression is required, the optical thickness of the layers A and B is given by

$$n_A t_A = n_B t_B = \lambda_1 \lambda_2 / 2(\lambda_1 + \lambda_2)$$

Since the total thickness of the period is $\lambda_0/2$, the thickness of layer C is given by

$$n_C t_C = \frac{\lambda_0}{2} - \frac{2\lambda_1\lambda_2}{(\lambda_1 + \lambda_2)}$$

Any two of the indices n_A, n_B, n_C can be chosen at will and the third calculated using the relation

$$\tan^2\left(\frac{\pi}{(1 + \lambda_1)/\lambda_2}\right) = \frac{n_A n_B - n_C^2}{n_B^2 - (n_A n_C^2/n_B)}$$

For example, for suppressing second- and third-order reflectance zones, each of the five layers of the period is $\lambda_0/10$ thick. The conditions on thicknesses and indices for suppressing second-, third-, and fourth-order reflectances are

$$n_A t_A = n_B t_B = \lambda_0/12 \quad \text{and} \quad n_C t_C = \lambda_0/6$$

$$n_B = (n_A n_C)^{1/2}$$

Figure 2.9 shows[16] the transmittance of a multilayer where the second, third, and fourth zones have been suppressed using this design. The steepness of the edge, in general, can be increased by increasing the number of periods, the ratio n_H/n_L, and the order of the filter.

2.4.2. Band-Pass Filters

2.4.2.1. AR Coatings

These coatings are one example of broad-band pass filters. Another kind of band-pass filter is a combination of long-wave pass and short-wave

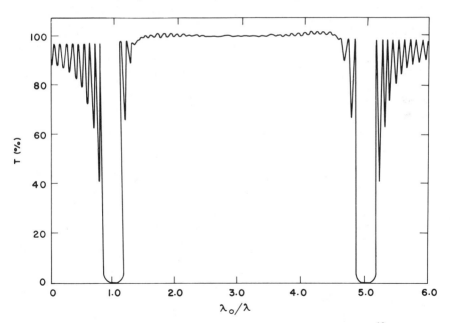

Figure 2.9. Calculated transmittance of the system substrate–A $(AB2C\,B\,A)^{10}$–air, with $n_s = 1.50, n_0 = 1.00, n_A = 1.38, n_B = 1.781, n_C = 2.30$, plotted as a function of λ_0/λ, showing suppression of second, third, and fourth reflection zones (after Ref. 16).

pass. This is a very useful arrangement for broad-band pass filters. Because of the difficulty of producing edge filters with sufficiently steep edges, bandwidths less than one-tenth of the central wavelength cannot be obtained.

2.4.2.2. *Fabry–Perot Filters (FP Filters)*

Design of narrow-band pass filters is based on the Fabry–Perot interferometer which consists of two identical highly reflecting optical flats spaced a distance d apart and aligned to be parallel. The transmittance of the FP assembly neglecting absorption and multiple reflections within the substrate is given by

$$T_F = \frac{T^2}{(1 - R)^2 + 4R \sin^2 \delta} \tag{2.24}$$

where R and T are the reflectance and transmittance of each reflector as seen from within the spacer layer and δ is given by

$$\delta = (2\pi/\lambda)nd \cos \Phi + \Psi \tag{2.25}$$

where n and d are, respectively, the refractive index and thickness, and Φ and Ψ are, respectively, the angle of refraction and phase change upon reflection in the spacer layer. A plot of Eq. (2.24) as T versus δ shows a series of narrow transmission bands centered on $\delta = m\pi$, where $m = 0, 1, 2, \ldots$ and is called the *order number* of the filter. Maxima in transmission occur at wavelengths given by

$$\lambda_0 = \frac{2nd \cos \Phi}{(m - \Psi)/\pi} \tag{2.26}$$

The peak transmittance T_0 is

$$T_0 = \left(\frac{T}{1 - R}\right)^2 = \left(\frac{1}{(1 + A)/T}\right)^2 \tag{2.27}$$

where $A = (1 - T - R)$ is the absorbance of each reflector. The minima in the transmittance value $T_{min} = (T/1 + R)^2$ occur for $m = 1/2, 3/2, 5/2, \ldots$. The rejection ratio defined as T_{min}/T_0 is, therefore, equal to $[(1 - R)/(1 + R)]^2$.

For $R > 0.7$, the half-width of the transmission band, expressed as a percent of λ_0, is given by

$$\frac{\Delta\lambda_H}{\lambda_0} \times 100 \approx \frac{1 - R}{\sqrt{R}} \frac{100}{2\pi nd \cos \Phi/[\lambda_0 + \Psi - \lambda_0(\partial\Psi/\partial\lambda)]} \tag{2.28}$$

which for the cases where dispersion in phase can be neglected, as for metal reflectors, reduces to

$$\frac{\Delta\lambda_H}{\lambda_0} \times 100 = \frac{1-R}{(R)^{1/2}} \frac{100}{m\pi} \qquad (2.29)$$

This shows that for a given λ_0, the higher the reflectance and the order of the filter, the narrower is the band.

A metal–dielectric filter represented as MDM is the simplest form of FP filter and consists of two metal reflectors on either side of a dielectric spacer. Equation (2.29) is valid for this case and shows that the half-width depends only on the reflectance of the metal layers and the order of the filter, typically being 1 to 8% with $T_0 \sim 20$ to 40%.

The typical characteristics of first-order (MDM) and second-order (MDDM) metal–dielectric filters for the visible region is shown[17] in Fig. 2.10. The short-wave sidebands of this filter can be suppressed by using a long-wave pass absorption filter. The filter can be designed to any wavelength by changing the thickness of the spacer layer. The commonly used materials in the visible region are Ag and Na_3AlF_6, and in the UV they are Al and MgF_2.

The performance of metal–dielectric filters is limited by absorption in the metal layers. Narrower filters with much higher peak transmission can

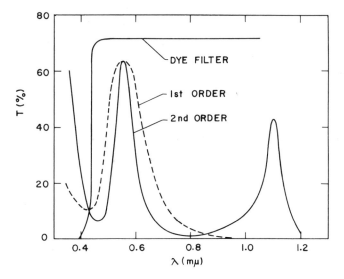

Figure 2.10. Characteristics of a first-order (MDM) and a second-order (MDDM) metal–dielectric filter. Also shown is the transmission curve of an absorption (dyed glass) filter used for the suppression of short-wavelength sidebands (after Ref. 17).

be constructed by replacing metal layers with all-dielectric reflectors described in Section 2.3.2. Typical design are $[(HL)^p\ 2H(LH)^p]$ and $[H(LH)^p\ 2L(HL)^pH]$. With such filters half-widths of the order of 1% with T_0 of the order of 80% can be readily achieved. The measured characteristic of an all-dielectric $(ZnS–MgF_2)$ FP filter of the type $[(HL)^3\ HLLH(LH)^3]$ in the visible region is shown[17] in Fig. 2.11. Since the all-dielectric reflector is effective over a limited range only, sidebands appear on either side of the peak and require suppression for most of the applications. The short-wave sidebands are removed by combining the FP filter with a long-wave pass absorption filter and long-wave sidebands with a metal dielectric filter used in the first order, as it does not have long-wave sidebands. This reduces the peak transmission of the filter to around 40 to 50%.

2.4.2.3. Multiple-Cavity Filters

Filters with more rectangular transmission peak and greater rejection in the stop band can be obtained by coupling two or more FP filters, each of which can be thought of as a resonant cavity. For example, $[(HL)^p\ 2H(LH)^p\ C(HL)^p\ 2H(LH)^p]$ represents a two-cavity or double-

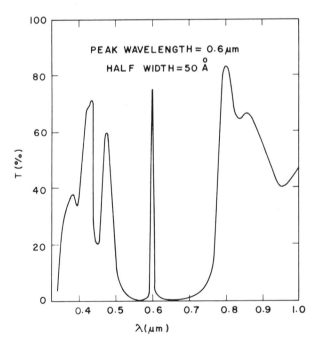

Figure 2.11. Measured transmission curve of a narrow-band all-dielectric $(ZnS–MgF_2)$ FP interference filter of the type $[(HL)^3\ HLLH(LH)^3]$ with unsuppressed sidebands (after Ref. 17).

half-wave (DHW) filter, where quarter-wave layer C is called the coupling layer. Some of the typical designs are: (DHW)HLLHLHLLH and HLHHLHLHLHHLH (due to Smith[18]); and triple-half-wave (THW)HHLHHLHH (due to Turner[19,20]). The advantage of multicavity filters over FP filters is their reduced sensitivity to residual absorption in the various layers.

2.4.2.4. Induced-Transmission Filter (FTR)

This is similar to the DHW all-dielectric filter in which the central reflectors and the coupling layer are replaced by a single metal layer. Designed by Berning and Turner,[21] this filter combines the wide rejection region of a metal–dielectric filter with peak transmittances that are closer to all-dielectric filters.

2.4.2.5. Frustrated Total Reflection (FTR) Filter

A high reflectance with no absorption and a minimum of scattering required for FP filter is possible using total reflection from a surface. Turner[22] devised such a system by sandwiching a spacer layer between the hypotenuse faces of two prisms coated with a lower-index layer, called the *frustrating layer*. The indices are chosen so that the critical angle is less than the angle of the prism and therefore light is incident at an angle greater than the critical angle. Due to the large angle of incidence at which the device must be used, an enormous shift in peak wavelength between the two planes of polarization appears. Billings[23] used a birefringent spacer layer to bring the two transmission bands together. A birefringent frustrating layer may also be used for the same purpose if suitable materials are available. FTR filters (also called optical tunnel filters) have not found wide applications because theoretically expected peak transmittances and half-widths could not be achieved and because the angular variation of the wavelength of the transmission peaks is very high.

2.5. Thin Film Polarizers

The high degree of polarization produced upon reflection at Brewster's angle has been exploited to obtain polarizers. Efficient large-aperture polarizers may be obtained in a simple fashion using thin film elements much smaller in number than in conventional pile-of-plates polarizers. By a suitable choice of the thickness and refractive index of the film, the reflectance of one of the polarized components is suppressed while that of the other is enhanced. Abeles[24] studied the properties of TiO_2 films on glass as polarizers. A glass plate coated on each face with TiO_2 of calculated thickness was found to transmit 92% polarized light. Whereas 12 uncoated glass plates are required to obtain 96% polarized light, only two TiO_2-coated plates are sufficient to give 99.7% polarized light.

A much higher degree of polarization with fewer elements can be achieved[25] with minimum loss and lateral beam displacement using multilayers of the form $(HL)^P H$. The higher the refractive index ratio, the fewer the layers required to achieve a certain degree of polarization, and the wider the spectral region over which the polarizer will act.

2.6. Beam Splitters

The use of thin metal films as neutral beam splitters has already been mentioned in Section 2.3.1. Now we describe two other types of beam splitters which do not suffer from the absorption losses accompanying a metal film.

2.6.1. Polarizing Beam Splitter

The principle of this polarizer is the same as described in Section 2.5. Construction differs from the ordinary multilayer $(HL)^P H$ polarizer in that the stack is embedded between two right-angled prisms. This type of beam splitter has a very large aperture and the transmitted beam is obtained without lateral displacement. It was first constructed by Banning[26] using ZnS–Cryolite–ZnS coatings on each prism and cementing them. An average polarization of 98% for white light incident within ±5° from the normal was obtained. Polarization varied from 99.7% in the green to a minimum of 94.5% in the blue.

2.6.2. Dichroic Beam Splitter

As the name implies, this is a device for splitting a beam into two colors. Dichroic beam splitters are essentially cut-off filters or occasionally band-pass filters designed for use at nonnormal angles of incidence, so that the reflected as well as the transmitted beam can be recovered. The larger the angle of incidence, the stronger is the dependence of spectral characteristics on the state of polarization of light. The error due to this can be reduced by (1) inserting a circular polarizer in the system, (2) making the high-index layer thicker than the low-index ones, and (3) using the stack in higher orders.

2.7 Integrated Optics

The rapidly advancing technology of thin films has led to the emergence of the exciting field of integrated optics,[27-29] which involves miniaturization of optical components. Since the wavelength of light waves is four to five orders shorter than that of microwaves, thin film optical components can be made very small and laid down on a common substrate with interconnections, forming an integrated unit of optical circuitry which is more compact, less vulnerable to vibrations and ambient temperature changes, and more economical as compared to the conventional bulk optical equipment aligned on heavy optical benches. Integrated circuits can be monolithic or hybrid, depending on whether various components like source, guide, modulator, detector, etc. are made out of the same material (single substrate) or different materials. To date, the only material which can perform all optical functions is GaAs, usually alloyed with Al.

The various optical devices discussed in the previous sections of this chapter pertain to light propagation across the film. In integrated optics light is made to propagate along the plane of the film. Structures that are used to confine and guide light in a particular direction are called *waveguides* and form the basic unit of all other optical components. Although the field of integrated optics has not yet reached maturity, some of the important potential applications which may find commercialization soon are light-wave communication and optical data processing. The importance of the use of optical frequencies in both these applications lies in the fact that (1) the information-carrying capacity of light waves is much greater than microwaves, and (2) optical circuits are much faster than electrical circuits. In what follows, we will briefly discuss the role of thin film technology in reaching the present state of the art in integrated optics.

2.7.1. Waveguides

The simplest waveguide is the two-dimensional planar slab guide consisting of a thin film of refractive index n_f sandwiched between a substrate and a cover material (usually air) of lower refractive indices n_s and n_c, respectively. Substrate and cover are together called *cladding*. The waveguide is said to be symmetric if $n_s = n_c$. To understand wave propagation in a guide, consider an asymmetric waveguide where $n_f > n_s > n_c$. Let θ_s and θ_c be the critical angles for total internal reflection from the film–substrate and film–cover interface, respectively. Since $\theta_s = \sin^{-1}(n_s/n_f)$ and $\theta_c = \sin^{-1}(n_c/n_f)$, obviously $\theta_c < \theta_s$. For very small angles of incidence (θ) in the film (i.e., $\theta < \theta_c < \theta_s$), light incident from the substrate side is refracted according to Snell's law and escapes from the guide through the cover (Fig. 2.12a). In this case, there is no confinement of light in the guide,

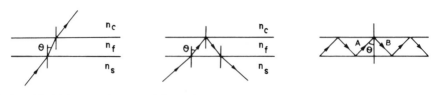

(a) RADIATION MODE   (b) SUBSTRATE RADIATION   (c) GUIDED MODE
                              MODE

Figure 2.12. Zigzag wave pictures of (a) radiation mode, (b) substrate radiation mode, and (c) guided mode in a thin film waveguide.

and the corresponding propagation mode is called *radiation mode*. For values of θ greater than θ_c but less than θ_s (i.e., $\theta_c < \theta < \theta_s$), the light incident from the substrate side after refraction at the film–substrate interface is totally reflected at the film–cover interface (Fig. 2.12b). Again there is no confinement of light, and one talks of *substrate radiation mode*. Finally, when θ is large enough so that $\theta_c < \theta_s < \theta$, the light is reflected back and forth in the film due to total internal reflection at both interfaces, and it is trapped, remaining confined to the plane of the film (Fig. 2.12c). The propagation mode corresponding to this is called *guided mode* and is physically pictured as a plane wave propagating along a zigzag path in the film. The zigzag wave motion is represented by two wave vectors A and B, each of magnitude kn_f, where $k = \omega/c$, ω being the angular frequency of light and c, the speed of light in vacuum. The horizontal components $kn_f \sin \theta$ of A and B determine the wave velocity parallel to the film. The vertical components $\pm kn_f \cos \theta$ determine the field distribution across the thickness of the film. Thus, in spite of the zigzag wave motion, the wave in a guided mode appears to propagate in the horizontal direction only; the vertical part of the wave motion simply forms a standing wave between the two film surfaces. The condition for a guided mode is given by

$$2kn_f t \cos \theta - 2\Phi_s - 2\Phi_c = 2m\pi \qquad (2.30)$$

where t is the thickness of the film, and $-2\Phi_s$ and $-2\Phi_c$ are the phase changes at the film–substrate and film–cover interfaces. The quantities Φ_s and Φ_c are given by the relation

$$\tan \Phi_{s,c} = \begin{cases} (n_f^2 \sin^2 \theta - n_{s,c}^2)^{1/2}/n_f \cos \theta, & \text{for TE waves} \\ (n_f^2/n_{s,c}^2)(n_f^2 \sin^2 \theta - n_{s,c}^2)^{1/2}/n_f \cos \theta, & \text{for TM waves} \end{cases}$$

$m = 0, 1, 2, \ldots$ is the order of the mode. Each mode is identified by the angle between the two legs of the zigzag. To excite a particular mode, the beam must enter the guide from a specific direction. From the dispersion relation, it can be shown that for an asymmetric guide, a certain cut-off thickness exists below which it cannot support a guided mode. For a symmetric guide ($n_c = n_s$) there is no cut-off, although the number of guided modes it can support increases as the thickness increases.

The planar guides discussed above provide no confinement of light in the plane of the film. Additional confinement can be obtained by using channel or strip guides (also called three-dimensional guides). Waveguides are fabricated using such techniques as vacuum evaporation, sputtering, ion implantation, diffusion, plasma polymerization, and LPE in combination with the standard techniques of pattern generation (lithography). Waveguides of some organic materials like PMMA and polystyrene are fabricated by solution spinning or dip casting. Some of the materials which have been studied for their use in construction of waveguides are sputtered Ta_2O_5, sputtered and epitaxial ZnO, evaporated ZnS, As_2S_3, and CeO_2, epitaxial GaAs, epitaxial garnets, and sputtered and epitaxial $LiNbO_3$, $LiTaO_3$, etc.

The various optical functions in integrated optics obviously cannot be performed by the conventional optical components used in three-dimensional (bulk) optics and therefore require the corresponding thin film equivalents, for example, thin film prisms, lenses, gratings, etc. These are described in the following section.

2.7.2. Thin Film Optical Components

The propagation of light in a two-dimensional medium, such as a light guide, can be described by an effective index of refraction N which is a function of the thickness t of the guiding film. The refractive index increases with increasing thickness. Therefore, a light beam in a guide is refracted or totally reflected at a step in the guide. Based on this principle, one can fabricate lenses, prisms, and similar elements by suitably shaping the thickness profile of the guide, either by deposition through masks or by controlled etching techniques. A thin film prism consists of a triangular region of higher thickness (Fig. 2.13a). When the region of higher thickness has curved surfaces, a thin film lens is obtained (Fig. 2.13b). Thin film lenses with f numbers as low as 2 to 3 have been fabricated by Schubert and Harris[30] by vacuum evaporation of CeO_2. It is also worth mentioning that whereas in bulk optics focusing is obtained by a convex lens only, in two-dimensional optics even a concave lens can focus light if its refractive index is made smaller than the surrounding region, for example, by making the lens region thinner. As shown in Fig. 2.13, the transition from higher- to lower-thickness regions is tapered to reduce the radiation losses at these boundaries. The effective index of refraction may also be modified by varying the composition of the guide.

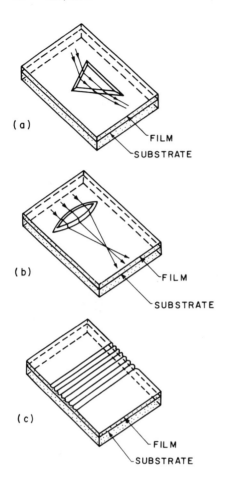

(a)

FILM
SUBSTRATE

(b)

FILM

SUBSTRATE

(c)

FILM   Figure 2.13. Thin-film prism, lens, and
SUBSTRATE   grating structures.

Another very important structure in two-dimensional optics is the periodic waveguide in which the variation in thickness is periodic along the direction of propagation. The propagation constant of a mode in such a structure undergoes periodic variation resulting in diffraction effects similar to a diffraction grating. These periodic structures are used in the fabrication of couplers, spectral filters, mode-selection devices, mirrors, etc., and are fabricated by the usual pattern-generation techniques described in Chapter 1. One of the common methods of fabricating a grating consists of developing a photoresist film exposed to the interference pattern obtained by splitting and recombining a laser beam. Depending on the photoresist and the developing procedure, the grating profile may assume a variety of shapes, e.g., sinusoidal, triangular, or rectangular. Gratings may also be formed by depositing through mask a low-index material on the guide. Thin film equivalents of amplitude masks and spatial filters are formed if the material deposited is highly absorbing. Spectral filters having stop bands and pass bands are fabricated using grating structures shown in Fig. 2.13c. Reflectance as high as 75% and a band width of 2 Å have been obtained by Flanders *et al.*[31] using a corrugated sputtered glass filter. A simple metal layer such as aluminium on top of a waveguide acts as a polarizer by absorbing TM modes and transmitting TE modes. Let us now see how these basic components are utilized in the fabrication of various devices for integrated optical circuits.

2.7.3. Passive Devices: Couplers

Devices which couple either a freely propagating beam into a guided mode or one guide to another are called *couplers*. Obviously, a beam cannot be coupled into a guided mode by simply having the laser beam incident on top of the film. Although a beam can be coupled to a guide by focusing it on the end face of the guide, good coupling efficiency is obtained only when the face is extremely well polished.

A beam can be efficiently (80% efficiency) coupled into or out of a guide using a prism coupler.[32] It consists of a right-angled prism of index $n_p > n_f$, placed on top of the guide (Fig. 2.14a) in such a way that the uniform air gap between the prism and the film is of the order of the

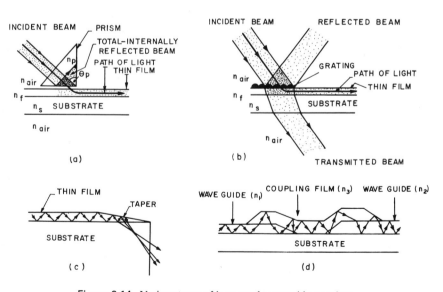

Figure 2.14. Various types of beam and waveguide couplers.

wavelength of light. Leakage of energy across the air gap occurs into the mth mode through the evanescent wave when

$$(\omega/c)n_{\rm p} \sin \theta_{\rm p} = k_x^{(m)} \qquad (2.31)$$

where $k_x^{(m)}$ is the propagation vector for the mth mode, and $\theta_{\rm p}$ is the angle of incidence of the beam in the prism and satisfies the condition

$$\theta_{\rm p} > \theta_{\rm c} = \sin^{-1} \left(\frac{n_{\rm air}}{n_{\rm p}} \right) \qquad (2.32)$$

Another way of coupling a beam into a guide is by using a grating as shown in Fig. 2.14b. The efficiency of a grating coupler[33] is smaller than that of a prism coupler because of losses by (1) transmission through the film as the grating, unlike the prism, does not operate under a total reflection regime, and by (2) reflection of higher-order diffracted beams if $d/\lambda_{\rm a}$ is not sufficiently small ($\lambda_{\rm a}$ is the wavelength in air). Nevertheless, because of their planar geometry and relatively easy fabrication using the latest etching techniques, gratings are the most promising couplers for future integrated-optics circuits.

A very simple coupler, more useful as an output coupler, is the tapered-film coupler[34] based on the cut-off property of an asymmetric waveguide. It consists of a film that tapers down into the substrate. As shown in Fig. 2.14c, a mode guided in the film, on reaching the taper, undergoes zigzag bounces with progressively decreasing angle $\theta_{\rm m}$ as the taper narrows down, until $\theta_{\rm m}$ becomes smaller than the critical angle, when most of the energy is transferred into an outgoing divergent beam. Although its efficiency may be as high as 70%, the high divergence (up to 20°) makes it less attractive. However, in the case of light detectors, where divergence is not objectionable, it can be used effectively.

Two guides of indices n_1 and n_2 in a circuit can be joined by a third overlapping film of intermediate index, as shown in Fig. 2.14d.

2.7.4. Active Devices

2.7.4.1. Modulators and Switches

Optical modulation is a process by which information is placed on a light wave. In the case of LED's and semiconductor lasers (see next section), direct amplitude modulation can be obtained by varying the diode current. However, even in applications where the modulation frequency obtained by this method is high enough (~GHz), external modulation is preferred because laser light has a tendency to oscillate, making the modulation hard to control. An external modulator operates by altering some detectable property, such as amplitude, frequency, phase or polarization of light, through a change in refractive index. The index may be changed by the application of an electric field, magnetic field, or an acoustic wave, the time variation of which represents the information to be conveyed. Switching is the operation by which the spatial location of a coherent wave is changed in response to an electric field, magnetic field, or an acoustic wave. A device may act both as a modulator and as a switch.

Electro-optic modulators utilize the linear electro-optic effect (Pockels effect) exhibited by highly anisotropic crystals like $LiNbO_3$, GaAs, ZnO, etc., and operate either as polarization modulators or amplitude modulators through the fundamental effect of phase modulation, which by itself is difficult to detect. The simplest electro-optic modulator consists of a channel waveguide with a parallel pair of electrodes alongside it. A linearly polarized beam, upon entering the guide, is split into two orthogonal polarization modes TE and TM due to the natural birefringence of the crystal. The application of an electric field across the guide through the electrodes changes the refractive index, and hence the phase velocity, differently for the two modes, resulting in a phase difference $\Delta\Phi$ between them when they leave the guide. This polarization-modulated wave can be converted into an amplitude-modulated one with the help of a polarizer.

Another type of modulator is the electro-optic grating modulator[35] where voltage applied between two sets of interleaved electrodes (Fig. 2.15a) on an electro-optic material (usually higher-index $LiNb_x Ta_{1-x} O_3$ on top of a $LiTaO_3$ substrate) generates a periodic refractive index variation or, in other words, a phase grating. Modulation occurs due to diffraction of the guided wave by the grating. This type of modulator has the advantage of directly giving an amplitude-modulated output without an additional polarizer and can also be used as a switch and beam deflector.

A phase grating may also be produced by an acoustic-wave through the photoelastic effect. Acoustic waves are produced by interdigital transducers deposited on top of the guide. Fabrication of the acoustic-optic modulator[36] requires the guide material to be piezoelectric.

Another electro-optic modulator and switch which has no bulk analog is based on the principle of directional coupling. It consists of two channel guides separated by a distance less than the wavelength of the propagating light.[37] The two guides are coupled through their evanescent waves, and, as in the case of coupled pendulums, the energy is exchanged between them periodically along the length of the guide, the period increasing with the increase in separation. Transfer of energy is complete only when k_x of the two guides is the same. Application of an electric field through the electrodes, shown in Fig. 2.15b, spoils the coupling and reduces the energy transfer. A voltage of 6 V is required to completely spoil the coupling

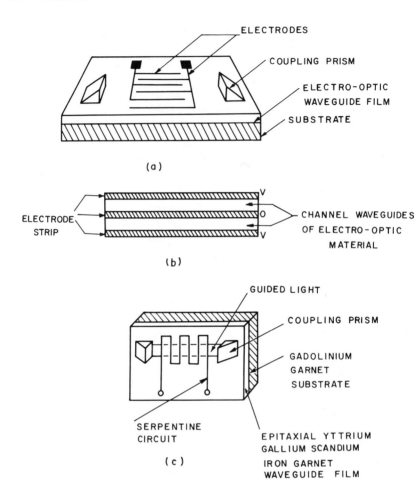

Figure 2.15. Schematic diagrams of the various types of optical modulators: (a) electro-optic; (b) directional-coupler; and (c) magneto-optic.

between two channel guides (width 2 μm, separation 3 μm) made by Ti diffusion into LiNbO$_3$.

A magneto-optical modulator[38] and switch is based on the principle of Faraday rotation. It consists of a magnetic iron–garnet film on a garnet substrate. Because of the lack of suitable magneto-optic materials transparent in the visible, this type of modulator is used only in the IR and near IR region. In operation, a dc bias field (magnetic) parallel to the waveguide plane and at 90° to the propagation axis is applied. This magnetizes the film, with the magnetization vector lying in the direction of the field. A linearly polarized beam from a He–Ne laser coupled to the guide through a prism coupler travels through the guide without any change in the state of polarization. Actuating the electrical circuit on the guide (Fig. 2.15c) with the signal current continuously rotates the plane of polarization of the beam as it travels along the guide, and after a certain coupling length, mode conversion takes place. Mode conversion is observed by coupling the light out of the guide using a rutile (TiO_2) prism coupler. A polarizer may be introduced before the prism if the modulator is to be used as a switch.

All the modulators described above use surface electrodes and therefore suffer from the loss of driving electric field in the space above the guide. An efficient use of modulation power can be made by guiding light in the depletion region of a reverse-biased *p–n* junction, as has been demonstrated[39] in GaP diodes for use in the fabrication of phase modulators.

2.7.4.2. Light Sources: Lasers

The most promising and widely tested light sources for integrated optics are those derived from the ternary $Ga_x Al_{1-x} As$ alloy system. A very important property of this system is that the lattice parameter varies only by 0.14% as x varies from 0 to 1; this reduces the losses due to nonradiative recombination centers introduced by lattice strain. By varying the alloy composition in the electrically active region (the recombination region), the emission wavelength can be varied over a useful range of 0.8–0.9 μm in which the attenuation and dispersion in the optical fibers is low. Before describing the actual laser structures, let us see how light can be guided in semiconductor layers. The first method consists of growing a highly resistive pure GaAs epitaxial layer on a low-resistivity *n*-GaAs substrate. Guiding takes place in the top layer because its refractive index is higher due to low free-carrier concentration. Another way of fabricating guides in GaAs is by diffusing, or ion-implanting, acceptor impurities into an *n*-type GaAs substrate to form a *p–n* junction. Light is guided in the depletion region between the *p*- and *n*-type regions because it has a higher refractive index than the surrounding regions, due to carrier compensation. In cases where the depletion region is too thin to confine light strongly, the higher-index *p*-layer may also contribute to guiding. High-index layers may also be obtained by varying the material composition; for example, addition of AlAs to GaAs reduces its refractive index. Therefore, the simplest guide using the $Ga_{1-x} Al_x As$ ternary system is the structure epi-GaAs–epi-GaAlAs–GaAs substrate. Another structure which, being symmetric, has the advantage of no cut-off for the fundamental mode and which is very useful in fabrication of lasers is the double heterostructure (DH) of the type GaAs–GaAlAs–GaAs–GaAlAs–GaAs substrate. Light is guided in the buried high-index GaAs layer. GaAlAs serves to isolate the guide from the outer GaAs layer.

Among the various possible configurations for continuous wave (CW) room-temperature laser diodes, the symmetrical DH[40] with a stripe contact (Fig. 2.16a) is the most suitable geometry for integrated optics. Lasing

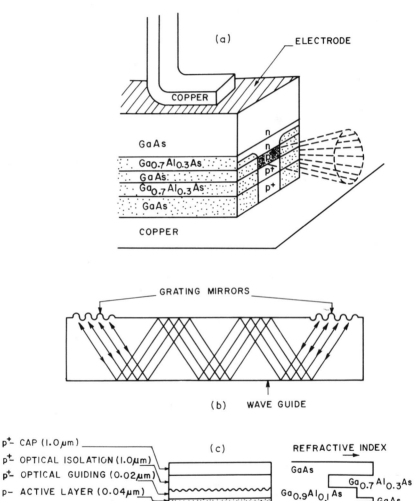

Figure 2.16. (a) Stripe-geometry double-heterostructure laser with dotted area as proton-bombarded region (after Ref. 40); (b) thin-film laser; (c) distributed-feedback double-hetero-structure laser and the corresponding refractive index profile (after Ref. 40).

action takes place in the GaAs layer at the p–n junction. Under forward bias conditions current is converted to light by radiative recombination of the injected electrons with holes in the active region. The emitted light is confined to the high-index p-GaAs by the lower-index $Ga_{0.7}Al_{0.3}As$ layers on either side. The dotted portion in the figure represents the proton-bombarded highly resistive region, so that current flows only in the stripe region from whose active region light is emitted. Typically the stripes are 17 μm wide with an active region 0.4 μm thick. Large wafers with grown layers are cleaved into 400-μm-long (total thickness $\sim$100 μm) dice to provide reflecting ($\sim$30%) end faces which are necessary for laser action. Power emitted at the end face of the crystal in the (17×0.4) μm^2 active region is typically 10–30 mW. Threshold current above which the lasing action takes place is typically $\sim$6000 A/cm^2. The laser dice is bonded with Sn with substrate up so that good thermal contact with the Cu heat sink is available for the active region.

Since the mirror surfaces for laser feedback are obtained by cleaving, this miniature laser dice cannot be used in monolithic integrated optic circuits. To make it compatible with monolithic applications the feedback must be provided by methods other than cleaving. A ridged-channel GaAs layer would suffice if the end walls of the ridge can be made flat enough to provide good mirrors. Such a GaAs mesa laser has been grown[41] by vapor-phase epitaxy on planar substrates through oxide masks. Diode (or injection) lasers have the advantage of directly converting electrical energy into light and therefore do not require an external source of light as in the case of optically pumped lasers.

Use of grating structures instead of mesa facets provides a completely new method of distributed feedback in a manner compatible with planar technology. An efficient mirror is formed of a grating with the optical length of the period equal to half the wavelength of the light to be reflected. In the case of an optically pumped laser, grating mirrors are placed at the two ends as shown in Fig. 2.16b. The position of the grating mirror in a DH injection laser for monolithic applications is shown[40] in Fig. 2.16c along with the corresponding refractive index profile.

2.7.4.3. Detectors

Detectors for monolithic integrated optical circuits make use of the photoconducting properties of semiconductor junctions and are of three types: (a) epitaxial photodetectors, (b) electroabsorption detectors, and (c) ion-implanted detectors. Each of these basically involves introduction of localized regions of lower-band-gap material into the higher-band-gap waveguide. In epitaxial photodetectors, holes 125 μm in diameter are made[42] by photolithography in a pyrolytically deposited SiO$_2$ layer on a

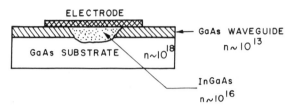

Figure 2.17. Integrated optical photodetector.

5-μm-thick, high-resistivity GaAs layer grown by vapor-phase epitaxy (VPE) on a lower-resistivity GaAs substrate. GaAs is removed from the hole regions by chemical etching, and an $In_x Ga_{1-x} As$ layer (with x varying from 0 to the required composition to reduce crystal imperfections due to strain) is epitaxially grown in the hole region. After smoothing the grown region by polishing, the SiO_2 mask is removed and Pt Schottky-barrier electrodes are evaporated, forming the complete detector (Fig. 2.17).

Electroabsorption detectors are based on the principle of electroab-sorption in which the lower-band-gap region is obtained by the application of an electric field. Electrically enhanced absorption for light near the band gap generates electron–hole pairs which are swept by the electric field, resulting in a photocurrent. A useful detector is made if the band gap of the electroabsorbing material is tailored to the wavelength of the light to be detected. This type of detector is a narrow-band detector and has the advantage of not requiring the formation of localized regions of lower-band-gap material. An electroabsorption detector has been[43] demonstrated in a reverse-biased *p–n* junction double heterostructure.

Other ways of decreasing the band gap and increasing absorption are ion implantation and impurity diffusion. A proton-implanted integrated-optics photodetector has been fabricated[44] in GaAs. In order that the impurity-diffused region may detect light, the impurity absorption must be a process for creation of free carriers.

3

Optoelectronic Applications

3.1. Introduction

This chapter describes applications based on photoelectronic effects which arise due to direct interaction of the incident photons with electrons in the material. These photon–electron interactions give rise to a variety of effects which can be exploited for device fabrication. It should be made clear here that these effects do not include the ones which arise due to heating of the material by absorption of photons and are therefore thermal or bolometric effects.

The most widely used photoeffects are those which involve conversion of incident optical energy to an electrical signal which can be measured to provide a quantitative indication of the incident photon flux. These are: (1) photoconductive, (2) photoemissive or photoelectric, and (3) photovoltaic. These effects form the basis of radiation detectors, picture tubes, electrophotography, and solar cells. Of equal importance in similar devices are the effects which involve conversion of electrical energy to optical energy. These effects are (1) cathodoluminescence, (2) electroluminescence, and (3) electrochromism, and they form the basis of various kinds of display devices. More recently, some very interesting photoeffects have been observed in amorphous chalcogenides, which hold potentialities for applications in a large number of areas such as computer memories, lithography, photography, etc. Some of these effects are: (1) photoinduced structural changes, (2) photo-induced optical changes, (3) photo-induced contraction, and (4) photo-induced chemical solubility changes. Optical information storage devices utilize a combination of the above-mentioned photoeffects. We prefer to classify the various devices according to their field of application and not according to their mode of operation.

3.2. Photon Detectors

In this section we describe only the photoconductive and photoemissive detectors. The photovoltaic detectors are dealt with in Section 3.3. Some of the general characteristics of the three types of photon detectors are:

1. A high speed of response as compared to thermal detectors.
2. A wavelength dependence of the photosignal per unit incident radiant power which increases with wavelength to a long-wavelength limit beyond which the photosignal falls to zero.
3. Detectors with long-wavelength limit in the UV, visible, or near IR (2–3 μm) operate uncooled at room temperature, whereas those with long-wavelength limit up to 4–5 μm and 8–14 μm require cooling down to dry ice temperature (195 K) and liquid N_2 temperature (77 K), respectively. This is necessary to reduce the noise due to the dark current produced by thermal energy.

3.2.1. Photoconductive Detectors

The interaction of photons having energy equal to or greater than the band gap with bound electrons of lattice atoms creates free-electron-hole pairs by excitation across the forbidden gap. In materials such as II–VI compounds, where the minority carriers are less mobile than the majority carriers, the presence of certain levels having much higher capture cross section for minority carriers than for majority carriers causes the minority carriers to be captured, leaving the majority carriers free for conduction. These levels may be due to defects, surface states, or any other mechanism responsible for immobilizing the minority carriers, thus making them unavailable for direct recombination. This effect of enhanced conductivity by absorption of radiation due to increased carrier concentration and/or lifetime of the photoexcited carriers is called *photoconductivity*.[1] The nature of defect centers is critical in determining the behavior of majority carriers and hence the characteristics of a photoconductor. In order to achieve high sensitivity, the important defect center should not only behave as a trapping centre in thermal equilibrium with the conduction or valence band, but must also behave as a recombination center and not be greatly affected by thermal equilibrium considerations. This is important because a highly photosensitive photoconductor should not only show a large change in conductivity upon illumination, but also respond fast. If only trapping centres are present, the response time of the photoconductor would be extremely long, which is not suitable for practical applications. If, on the other hand, recombination centers predominate, the response time is determined by the free-carrier lifetime. However, at low excitation intensities, the speed of response is orders of magnitude slower because the initial small density of photoexcited carriers is rapidly trapped. Trapping also lengthens the decay time at all levels of excitation, since the carriers are slowly released after removal of the excitation source. Enhancement in the photosensitivity of a material is achieved by deliberately introducing suitable impurities into the material. In CdS, for example, which is of n type, introduction of either donor or acceptor impurity may enhance the photosensitivity. Iodine acts as a donor in CdS and generates acceptor-like cation vacancies in an attempt to achieve charge compensation, which take up a negative charge with respect to the cation sublattice and therefore have a higher capture cross section for photoexcited holes than for photoexcited electrons. Similarly, Cu, an acceptor in CdS, sensitizes it by forming a negatively charged copper center on the cation sublattice.

A photoconductive detector simply consists of a thin film of a photoconductor with electrodes depositied on top of it, as shown in Fig. 3.1. The whole unit is encapsulated to protect it from environmental effects such as dust and humidity. The commonly used detectors in the visible region are vacuum-evaporated or spray-pyrolyzed thin films of CdS, CdSe, and $CdSe_{1-x}S_x$. In the infrared region solution-grown films of PbS, PbSe, and $PbSe_{1-x}S_x$ are commonly used for fabrication of photodetectors. We have shown[2-4] in our laboratory that solution-grown CdS, CdSe, and $CdSe_{1-x}S_x$ films photodetector fabrication in the visible region, and $Pb_{1-x}Hg_xS$ films are suitable in the infrared region.

A photoconductive detector is characterized by its gain, speed of response, and spectral response. A convenient way of defining the gain is

$$G = \text{light current/dark current}$$

Another definition of G is

$$G = \tau/T_R$$

where τ is the free-carrier lifetime and T_R is the transit time of the carriers across the electrodes. Comb-shaped electrodes as shown in Fig. 3.1 are

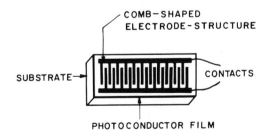

Figure 3.1. Schematic diagram of a photoconductive detector showing the comb-shaped electrodes.

chosen to provide maximum exposed area of the film and minimum transit time, so that the gain is maximum.

If one assumes a constant quantum efficiency and uses a constant incident photon flux, the detector response should remain essentially wavelength independent until it drops to zero when photons do not have sufficient energy to excite carriers. Since the spectral measurements are generally made for constant energy rather than constant photon flux, this behavior is not observed. Secondly, at high energies, the absorption constant is much larger with the result that only surface regions are excited where defect states give shorter lifetimes and therefore lower photosensitivity. A peak in the response curve occurs approximately at the wavelength for which the absorption constant is equal to the reciprocal of the film thickness. This generally corresponds to wavelengths near the absorption edge of the material. The presence of certain impurity states may extend the spectral response to longer wavelengths because of direct excitation of carriers from defect levels into the conduction band.

3.2.2. Photoemissive Detectors

As the name implies, *photoemission*[5] is a process in which the incident photons knock the electrons out of a material, called the *photocathode*. On a solid surface, the valence electrons are prevented from escaping by a potential barrier. The work function Φ of the material, which is defined as the difference in energy between the vacuum level and the Fermi level, gives a measure of the strength of this surface barrier. This definition of work function is illustrated in Fig. 3.2a and b for a metal and a semiconductor. For a metal at $T = 0$, Φ corresponds to the threshold energy for photoemission. At a finite temperature the spread in the Fermi distribution also allows photoemission for energies less than Φ. In a semiconductor or insulator, since the Fermi level lies within the forbidden gap E_g, the photoemission threshold E_T is given by $E_T = E_g + E_A$, where E_A is the electron affinity, defined as the vacuum energy level as measured from the bottom of the conduction band.

Unlike the photoconductive detector, a thin film of a photoemissive material by itself does not make the complete detector but only the photocathode. A photemissive detector, in addition to the photocathode, consists of an anode to collect the electrons and a number of dynodes (thin films of materials having good secondary emission characteristics) to provide amplification before collection, all enclosed in a tube with a window of suitable material to provide optical access. The photocathode is the most important part of the detector because its characteristics are the main factors determining the performance of the detector. Photocathodes are fabricated using thin film deposition techniques, usually vacuum evapor-

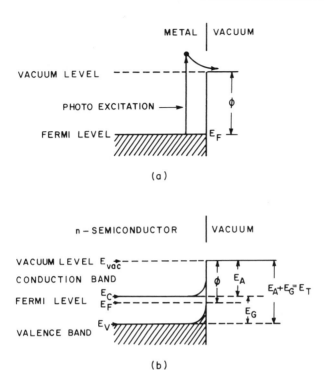

Figure 3.2. Diagram showing the work function of a metal and a semiconductor.

ation, except for photocathodes based on III–V compounds. Table 3.1 shows some of the commonly used photocathodes and their long-wavelength cutoff, which is the wavelength corresponding to the threshold energy and is a characteristic of the photocathode material. It can be seen from the table that metal photocathodes are not used in practice, not even cesium (Cs) which has the lowest work function among metals. Instead, semiconductors are used as photocathode materials. The reason for this becomes clear later on. Most of these semiconductors contain at least one element of low work function such as the alkali metals, Cs being the most common. These semiconductors are difficult to evaporate directly because of their tendency to decompose into individual elements, and therefore the fabrication of such photocathodes involves a two-step process: (1) evaporation of the nonalkali metal, for example, Sb in Cs_3Sb, and (2) reaction of this film with the vapors of the alkali metal at temperatures ~ 150–200°C. Some of the materials shown in the table, for example, Cs_3Sb, KCl, and LiF, also possess desirable secondary emission characteristics and can, therefore, be used for dynode fabrication.

Table 3.1. Some Commonly Used Photocathodes, Their
Long Wavelength Cutoffs, and the Method of Fabrication

Photocathode	λ_0 (μm)	Method of fabrication
Cs_3Sb	0.65	VE^b
$(Cs)Na_2KSb$	0.83	VE
K_2CsSb	0.66	VE
K_2CsSb (surface oxidized)	0.78	VE
Cs_3Bi	0.70	VE
$Ag–Bi–O–Cs^a$		
($Cs_3Bi + Cs_2O + Ag$)	0.76	VE
Cs_2Te	0.35	VE
CsI	0.19	VE
KBr, KCl	0.15, 0.14	VE
LiF	0.12	VE
$Ag–O–Cs^a$	1.0	VE
$GaP/Ga_{0.25}Al_{0.75}As/GaAs$		
($Cs:Cs_2O$)	1.3	LPE/VPE
$GaAs(Cs:Cs_2O)$	0.92	CVD, VE

[a] The Ag layer in some cases has also been deposited by chemical reduction
plating.
[b] VE is vacuum evaporation.

Other than the conventional photocathodes, there exist two more types
of photocathodes having higher efficiences. These are: (1) intereference
photocathodes, and (2) negative-electron-affinity (NEA) photocathodes.
These are described in the following sections.

3.2.2.1. Interference Photocathodes

Highly photosensitive photocathodes with high quantum yield (i.e.,
number of photoelectrons/number of incident photons) must meet three
requirements:

(i) low threshold energy for emission;
(ii) no reflection loss of the incident radiation at the surface of the
photocathode; and
(iii) complete absorption of the photons within the attenuation length
of the photoelectrons—the greatest depth from which the photoexcited electrons can escape.

In a metal photocathode, because of limited effective absorption due
to high reflectance of metal surfaces, the overall efficiency is too low to be
of any use in practical applications. On the other hand, in a semiconductor

photocathode, although a sufficient number of photons enter the cathode (low reflection losses), the efficiency is limited because the absorption coefficient is small and the penetration depth of photons is much greater than the attenuation length of the photoelectrons. Therefore, only a fraction of photons is absorbed within the region from where emission can take place. Thus, in a semiconductor photocathode, a part of the incident light is lost by reflection and a part by transmission, but the overall efficiency is much greater than that of a metal photocathode. A semiconductor photocathode can be used both in the reflection mode (i.e. light incident from the cathode side), and in the transmission mode (i.e., light incident from the substrate side). The main function of an interference photocathode[6,7] is to enhance the efficiency of a photocathode exploiting optical interference techniques. When a semiconductor is used in the transmission mode, part of the incident radiation is lost at the substrate–photocathode interface due to different optical admittance values. This loss can be reduced by inserting a dielectric layer of a suitable refractive index between the substrate and the photocathode such that the total optical thickness of the dielectric + photoelectric is $\lambda/2$, where λ is the wavelength corresponding to the incident photons. The dielectric layer acts as an antireflection coating for the photocathode. Sometimes two layers may be added, one for phase matching and the other for amplitude matching. Such photocathodes are respectively called *single-layer* and *double-layer transmissive interference cathodes* (TIC). By using TiO_2 as the matching layer, the quantum efficiency of a Cs_3Sb photocathode can be increased by a factor of 1.5. The increased efficiency of a TIC is also accompanied by an increased transmission. The efficiency of a TIC (Fig. 3.3a) can be increased further by a factor of 1.5 by illuminating it at an angle greater than the critical angle for the photoelectric–air interface. In such a case, the photocathode is called a *totally reflective interference cathode* (TRIC). This is a very useful arrangement in those cases where the photocathode is very thin (thickness less than the emission depth of the photoelectrons), and more so if the incident light is monochromatic and plane-polarized.

The oblique incidence in a TRIC severely restricts its application. Another type of cathode operated in the reflective mode and which can be used at all angles of incidence is the *reflective interference cathode* (RIC). The basic arrangement of an RIC is shown in Fig. 3.3b. Here, radiation is incident from the photoelectric layer side. To reduce the transmission losses, a mirror such as an aluminum layer is placed between the substrate-tuning layer and the substrate, thus completely closing the cavity. For a multialkali photocathode, by using SiO_2 as the matching layer and silver as the mirror, the quantum yield can be increased[8] by a factor of 3.7. The RIC is the most efficient type of interference photocathode, with an efficiency enhancement factor of up to 8 in some cases.

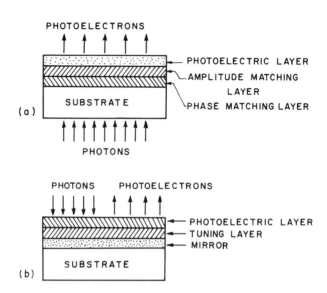

Figure 3.3. Interference photocathodes of (a) transmission type (TIC) and (b) reflection type (RIC).

3.2.2.2. NEA Photocathodes

The negative-electron-affinity photocathode is the most recent and the highest-performance photocathode. Negative electron affinity is obtained when the energy of an electron at the bottom of the conduction band in the bulk of a semiconductor is greater than the zero energy level of an electron in vacuum. Such a state can be obtained in a p-type semiconductor with a very thin overcoating of a low-work-function metal, usually deposited in the presence of oxygen. As a result of NEA, an electron excited to the conduction band within the bulk can, if it travels to the activated surface in the excited state, energetically fall out of the photocathode into free space. The concept of NEA can be made clear with the band diagram of such a photocathode (Fig. 3.4a). In a p-type material, band bending at the surface due to pinning by surface states at E_{ss} is downwards and assists in attaining of NEA, which cannot be attained in a n-type material because the upward band bending assists confinement of the electrons to the bulk. Band bending, as in a p-type material, is essential because it allows the cold electrons (i.e., "thermalized" to the bottom of the conduction band) to overcome any small positive barrier which might be present at the coated semiconductor surface. As illustrated in Fig. 3.4a, when the band bending E_{BB} is greater than this barrier E_B, zero or negative effective bulk electron affinity is obtained in spite of this

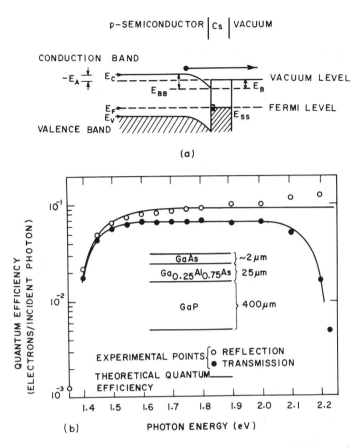

Figure 3.4. (a) Energy band diagram of a NEA photocathode; (b) characteristics of a multilayer GaAs photocathode (after Ref. 10).

barrier. Some of the important advantages of NEA photocathodes over the conventional ones are: (1) enhanced quantum efficiency, (2) extension of the response into the near IR region of the spectrum, where no conventional photocathodes with quantum efficiency greater than 0.1% are available, (3) possibility of extremely good resolution in image sensors because of the narrow energy distribution of the emitted cold electrons as compared to that of the hot electrons in the conventional photocathodes, (4) extremely low room-temperature dark current even for surfaces sensitive beyond 1.1 μm, and (5) extremely uniform response over an extended region of the spectrum. NEA photocathodes using Cs–O activated GaAs prepared by CVD on glass or sapphire substrates have been fabricated[9] with performance comparable to that of the single-crystal photocathodes. Most of the

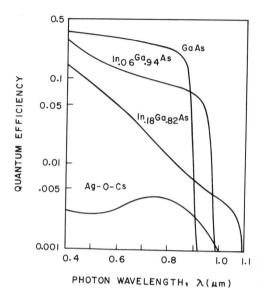

Figure 3.5. Characteristics of some NEA photocathodes. The S-1 AgCsO curve is for reference (after Ref. 11).

NEA photocathodes are based on III–V compounds. Figure 3.4b illustrates a multilayer GaAs photocathode[10] along with its performance. Layers of III–V compounds are grown by VPE or LPE (vapor/liquid phase epitaxy). The relative performance of some of the NEA photocathodes as compared to the conventional Ag–Cs–O photocathode, is shown[11] in Fig. 3.5.

3.3. Photovoltaic Devices

The photovoltaic effect arises when electron–hole pairs generated by the absorption of the photons (of energy greater than or equal to the band gap of the material) are separated by a built-in electric field due to an internal potential barrier, such as that existing at a *p–n* junction or a Schottky barrier. A photovoltaic junction, therefore, exhibits an open-circuit voltage V_{oc} and a short circuit current I_{sc} upon illumination and can be used as a means to convert sunlight into electrical energy. Such a device is known as a *solar cell*.

Knowledge of the salient features of a junction is essential in understanding the principles underlying energy conversion by a solar cell. For a detailed analysis of junctions, the reader is referred to standard textbooks[12–14] on the subject.

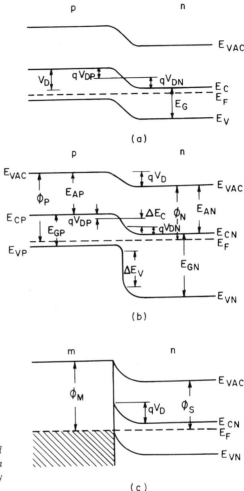

Figure 3.6. Energy band diagrams of (a) *p–n* homojunction, (b) a *p–n* heterojunction, and (c) a Schottky barrier.

3.3.1. *Solar Cells: General Analysis*

The energy-band diagrams of a *p–n* homojunction, a *p–n* heterojunction, and a Schottky barrier are shown in Fig. 3.6. Formation of the potential barrier in each one is clear from the energy-band diagrams. For understanding how power is generated in a solar cell, let us consider the simple case of a *p–n* junction. The space-charge region at the junction (also known as the depletion region) which gives rise to the built-in field, is formed by the transport of some holes from the *p* to the *n* region and of some electrons from the *n* to the *p* region. The direction of this field is such as to assist the transport of minority carriers and oppose the transport of majority carriers across the junction. Upon illumination, the semiconductor absorbs the solar radiation, generating electron–hole pairs according to

$$G(x) = \int_{E_g}^{\infty} \Phi(E)[1 - R(E)]\alpha(E) \exp[-\alpha(E)x] \, dE \qquad (3.1)$$

where $G(x)$ is the number of pairs generated at any point x (with the origin at the surface through which light enters), E_g is the band gap of the semiconductor, Φ is the photon flux at energy E, $R(E)$ is the reflectance at energy E, and $\alpha(E)$ is the absorption coefficient at energy E. The photogenerated minority carriers reach the junction and cross over into the regions where they are the majority carriers. These charge carriers flow through the external load resistance R, causing a potential drop V across it which makes the junction forward biased as shown in Fig. 3.7a and results in a dark current which opposes the light-generated current I_L. In order for a substantial fraction of the generated carriers to reach the junction, the carriers must be generated within a distance of the order of the diffusion length from the junction. This requirement imposes a dual condition on the material properties, viz. a high absorption coefficient and, consequently, a low absorption thickness, and a long diffusion length. Alternatively, a drift field can be provided in the bulk of the absorber which increases the effective diffusion length of the carriers. The drift field can be obtained either by a spatial gradient in the impurity concentration profile or by a spatially varying band gap in a graded composition material. It should also be noted that at the surface of a semiconductor, a large density of surface states exists. This leads to a large surface recombination velocity with the result that the carriers generated near the free surface diffuse toward this surface and are lost via surface recombination. A counteracting field, called the back-surface field or a minority-carrier-reflecting field, at this surface helps to direct the minority carriers away from this surface toward the collecting junction. Such a surface field is achieved by changing the doping profile at the surface, as in the case of Si and $CuInSe_2$–CdS solar cells, or by providing a heteroface junction at the surface, as in the case of GaAlAs–p-GaAs-n-GaAs and Cu_2O–Cu_2S–CdS solar cells. In both the cases, multi-layered structures are necessary.

The current I flowing through R is given by

$$I = I_L - I_D \qquad (3.2)$$

where I_D represents the diode current in dark. The corresponding current density, obtained by dividing I_D by the junction area, is given by

$$J_D = J_S[\exp(eV/kT) - 1] \qquad (3.3)$$

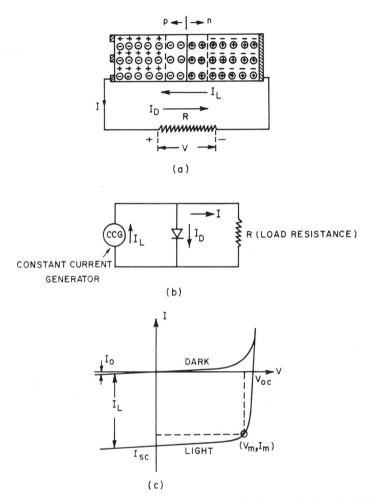

Figure 3.7. A p–n homojunction solar cell (a) along with its ideal equivalent circuit (b) and ideal I–V characteristics (c).

where e is the electronic charge, k is the Boltzmann constant, T is the absolute temperature, and J_S is the reverse saturation current density given by

$$J_S = e\left(\frac{D_n}{\sqrt{D_n \tau_n}} n_{0p} + \frac{D_p}{\sqrt{D_p \tau_p}} p_{0n}\right) \tag{3.4}$$

where D_n are D_p are the diffusion coefficients, τ_n and τ_p are the lifetimes and n_{0p} and p_{0n} are the equilibrium concentrations of electrons in the p

region and of holes in the n region, respectively. The quantities n_{0p} and p_{0n} are given by

$$n_{0p} = n_i^2/N_A \quad \text{and} \quad p_{0n} = n_i^2/N_D \tag{3.5}$$

where N_A and N_D are the acceptor and donor densities in the p and n regions, respectively, and n_i is the density of electrons (or holes) in the intrinsic material and is given by

$$n_i = KT^{3/2} \exp\left(-E_g/2kT\right) \tag{3.6}$$

Here, K is a constant and E_g is the band gap of the material.

The I–V characteristic of an ideal solar cell, represented by the equivalent circuit shown in Fig. 3.7b, is described by the equation

$$I = I_L - I_S[\exp\left(eV/kT\right) - 1] \tag{3.7}$$

where I_S is the reverse saturation current. A plot of this equation is shown in Fig. 3.7c. Equation (3.7) shows that the open-circuit (i.e., $I = 0$) voltage

$$V_{oc} = (kT/e) \ln\left[I_L/(I_S + 1)\right] \tag{3.8}$$

and the short-circuit (i.e., $V = 0$) current I_{sc} is given by

$$I_{sc} = I_L \tag{3.9}$$

It is convenient to define a parameter, called the fill factor (FF), which relates the maximum power a cell can deliver to V_{oc} and I_{sc}, by

$$\text{FF} = I_m V_m / V_{oc} I_{sc} \tag{3.10}$$

where I_m and V_m are the current and voltage corresponding to the maximum power P_m. The efficiency η of a solar cell is defined as

$$\eta = P_m/P_0 = (V_{oc} \times I_{sc} \times \text{FF})/P_0 \tag{3.11}$$

where P_0 is the input power (or the intensity of the incident radiation). Let us now discuss the various factors which determine V_{oc}, I_{sc}, and FF and hence the efficiency η of a cell.

3.3.1.1. Open-Circuit Voltage V_{oc}

It follows from Eqs. (3.4)–(3.6) and (3.8) that V_{oc} of a solar cell increases with:

(i) a decrease in the temperature of the junction;
(ii) an increase in the band gap of the material;

(iii) a decrease in the resistivity (i.e., an increase in N_A and N_D) of the material;

(iv) an increase in the minority carrier lifetime, and therefore the diffusion lengths $[L_n = (D_n\tau_n)^{1/2}, L_p = (D_p\tau_p)^{1/2}]$ in the material; and

(v) an increase in $I_L(= I_{sc})$.

3.3.1.2. Short-Circuit Current I_{sc}

We have already seen that $I_{sc} = I_L$. It can be shown[15] that

$$I_L = (p_0\alpha q/h\nu)e(L_n + L_p) \tag{3.12}$$

where α is the optical absorption coefficient of the material for photons of energy $h\gamma$, and q is the quantum efficiency, defined as,

$$q = \frac{\text{Number of electron–hole pairs generated}}{\text{Number of photons absorbed}}$$

In general, however, when a spectrum of radiation is incident on the cell, the quantity $p_0\alpha q/h\nu$ must be replaced by its integral over the entire spectrum.

It can be seen from Eq. (3.12) that I_{sc} increases with increase in minority carrier lifetimes and the diffusion lengths.

3.3.1.3. Fill Factor FF

It can be shown[16] that under ideal conditions

$$\text{FF} = \frac{V_m}{V_{oc}}\left(1 - \frac{\exp(eV_m/kT) - 1}{\exp(eV_{oc}/kT) - 1}\right) \tag{3.13}$$

We see that the FF depends only on V_{oc}, and not on I_{sc}. An ideal p–n junction solar cell with the I–V curve shown in Fig. 3.7(c) has a FF of 0.82.

The efficiency of an ideal p–n-homojunction solar cell for a given material of band gap E_g, as calculated by Landsberg,[17] is

$$\eta_u = \frac{E_g}{kT_s} \int_{E_g/kT_s}^{\infty} \frac{x^2\, dx}{e^x - 1} \bigg/ \int_0^{\infty} \frac{x^3\, dx}{e^x - 1} \tag{3.14}$$

where T_s is the blackbody temperature of the radiation source. A plot of the ultimate efficiency η_u with the E_g of the material, for the case of a perfect homojunction, is shown by curve I in Fig. 3.8. Curve II in Fig. 3.8 shows the theoretically calculated values of attainable efficiencies with

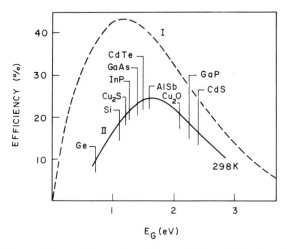

Figure 3.8. Ultimate efficiency (curve I) and theoretically calculated maximum practical efficiency (curve II) as a function of energy band gap.

various materials. However, the practical efficiencies obtained in real solar cells are much smaller that these theoretical values.

A real solar cell differs from an ideal one in having finite values of series resistance R_s and shunt resistance R_{sh}, which in the ideal case are zero and infinity, respectively. Correspondingly, the equivalent circuit is modified as shown in Fig. 3.9a.

The series resistance R_s consists of the resistance of the top and the base layers and the top and the bottom contacts. The shunt resistance arises due to various leakage paths. An increase in R_s reduces I_{sc} but does not affect V_{oc}. The open-circuit voltage V_{oc} is reduced first slowly and then rapidly with decrease in R_{sh}. The effect of R_s and R_{sh} on the FF is shown in Fig. 3.9b. The corresponding effect on the shape of the $I-V$ characteristics is shown in Fig. 3.9c.

Another reason for lower practical efficiencies is that the actual junctions are far from ideal, and therefore the FF is much lower than the theoretical value even if $R_s = 0$ and $R_{sh} = \infty$. The efficiency of a real solar cell is further lowered due to the reflection losses from the top layer. These losses can be drastically reduced by the use of AR coatings at the top (see Section 2.2).

Analysis of a $p-n$ heterojunction is similar to that for a homojunction, except for the fact that the band gaps of semiconductors on either side of the junction are different. Unlike the $p-n$ homojunction, the shortwave response of a $p-n$ heterojunction can be improved by using a wide-band-gap material for the front region which acts as a transparent window for short wavelengths. It should be noted that the efficiency of a heterojunction

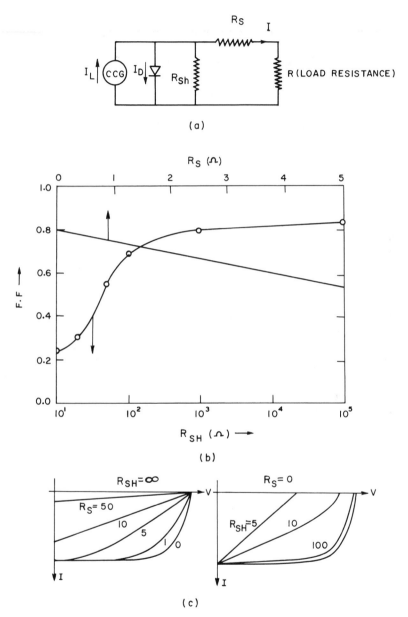

Figure 3.9. (a) Equivalent circuit of a real solar cell; (b) variation of the FF with R_s and R_{sh}; (c) effect of R_s and R_{sh} on I–V characteristics.

device is always lower than that of a homojunction device fabricated from the lower-band-gap material. For obtaining high efficiencies in the case of heterojunction solar cells, electron-affinity matching and lattice-parameter matching between the two materials are essential. This is so because the presence of a mismatch in the electron affinities gives rise to a notch, spike, or a discontinuity at the conduction band edge. A notch or spike would impede the flow of carriers across the junction, while a discontinuity would help in carrier recombination at the interface states. On the other hand, the presence of a mismatch in the lattice parameters of the two materials gives rise to a large density of states at the interface, providing a path for the recombination of carriers during their traversal across the junction. The stringent requirements of lattice matching and electron-affinity matching in heterojunction devices restrict the photovoltaically useful material combinations to CdS–InP, CdS–CuInSe$_2$, Cu$_2$S–CdZnS, and CdTe–CdZnS. To utilize more semiconductor materials, it is necessary to synthesize new ternary, quarternary, and quinary alloys with tailored band gaps, lattice parameters, and electron affinities.

In Schottky-barrier cells, higher values of V_{oc} are obtained by inserting an oxide layer between the metal and the semiconductor. Such solar cells are called, *metal–oxide–semiconductor* (MOS), *or metal–insulator–semiconductor* (MIS) solar cells. In the case of an MIS-type device, for optimum operation the band at the junction should be strongly inverted and the insulator layer should be thin enough so as to be transparent to the electrons crossing over from the semiconductor to the metal. The work function of the metal and the electron affinity of the semiconductor have to be properly chosen so as to obtain conditions of inversion at the interface. The thickness of the insulator layer, its coherence and integrity, and the surface states critically determine the performance of the device.

Excess energy and the energy of the photons incapable of creating electron–hole pairs account for the loss of 75% of the incident radiation in a conventional solar cell. Recent studies aimed at reducing these losses and hence improving the efficiencies include; (1) the optical-filter-mirror approach,[18,19] in which the incident solar radiation is split into parts by dielectric-filter mirrors (see Chapter 2) and directed onto separate solar cells each with spectral response matched to the radiation incident on it. A combined efficiency of 35–37% can be obtained,[20] for example, in GaAs–Si and GaAs–Ge combinations; (2) the tandem-cell approach,[21,22] in which a number of solar cells responding to different parts of the solar spectrum are placed one behind the other in such a sequence that the radiation is incident on the one responding to the highest energy photons, which is backed by successively decreasing maximum-response energies. High-energy photons are absorbed by the first cell, and the rest of the spectrum is transmitted so as to be incident on the second cell, which now

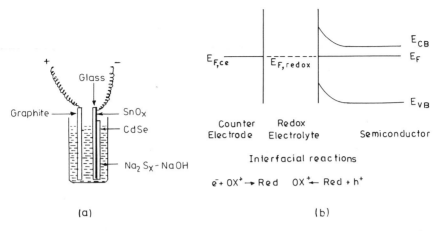

(a)   (b)

Figure 3.10. (a) Schematic diagram of a PEC cell and (b) the energy band diagram of a semiconductor–electrolyte junction.

absorbs the higher-energy photons of the transmitted spectrum and allows the rest of the spectrum to fall on the third cell, and so on. In this manner, the spectrum splitting is effected without the use of optical-filter mirrors. By using a large number of cells with band gaps separated by small amounts, an almost perfect match to the solar spectrum, in principle, can be obtained. The cells can be connected in series to obtain a high value of V_{oc}. However, for obtaining higher efficiencies, it is important that the cells be matched, i.e., equal photocurrents should flow through all the cells, otherwise power losses would occur.

A brief mention may also be made here of photoelectrochemical (PEC) cells, which are still in the developmental stage. A PEC cell consists simply of a semiconductor–liquid electrolyte heterojunction (shown in Fig. 3.10a) which behaves as a Schottky barrier with energy band diagram shown in Fig. 3.10b. The electrolyte must be capable of both reduction and oxidation, i.e., it must contain a redox couple. An example of the materials used and the efficiencies obtained can be seen in Table 3.2. The main problems associated with these easily fabricated PECs are packaging and the degradation associated with photodecomposition of the semiconductor electrodes. One very attractive application of PECs, if developed for long-term use, would be as combined solar-thermal and solar-electric generators, in which heat output is provided by the electrolyte (which becomes hot due to absorption of a major portion of the solar energy as heat) through a heat exchanger. Further, if the pump is driven by the electrical output of the cells itself, an interesting self-controlled system can be devised. For a detailed analysis of semiconductor–electrolyte junctions and the chemical

Table 3.2. Characteristics of Various PEC Cells[a]

Semiconductor (morphology), electrolyte	Deposition technique	V_{0c} (V)	J_{sc} (mA cm^{-2})	FF	η(%)
n-GaAs (SC), 0.8 M Se^{-2}, 0.1 M Se$_2^{-2}$, 1 M OH$^-$		0.72	24	0.70	12
n-GaAs (PC), 0.8 M Se^{-2}, 0.1 M Se$_2^{-2}$, 1 M OH$^-$	CVD	0.57–0.62	22	0.42	4.8
n-CdS (SC), 0.2 M Fe(CN)$_6^{-4}$, 0.1 M Fe(CN)$_6^{-3}$, 0.4 M KCl		0.95	6.2	0.68	5.5
n-CdS (PC), 1.2 M OH$^-$ 0.2 M S^{-2}	Spray pyrolysis	0.60	1.8×10^{-3}	0.21	0.038
n-CdS (PC), 1.25 M, 0.2 M S^{-2}	Evaporated	0.71	1.1×10^{-4}	0.25	0.054
n-CdTe (SC), 1 M Se^{-2}, 0.1 M Se$_2^{-2}$, 1 M OH$^-$		0.81	18.1	0.40	8.4
n-CdSe (SC), 1 M S^{-2}, 1 M S, 1 M OH$^-$		0.72	14.0	0.60	8.4
n-CdSe (PC), 2.5 M S^{-2}, 1 M S, 1 M OH$^-$	Coevaporated	0.51	8.2	0.61	5.0
n-CdSe (PC), 1 M S^{-2}, 1 M S, 1 M OH$^-$	Electro- deposition	0.52	9.2	0.55	3.1
n-CdSe (PC), 1 M S^{-2}, 1 M S, 1 M OH$^-$	Spray pyrolysis	0.63	11.8	0.35	5.2
n-CdSe (PC), 1 M S^{-2}, 1 M S, 1 M OH$^-$	Solution growth	0.41–0.62	12.0	0.45	2.25
n-WSe$_2$ (SC), 0.5 M K$_2$SO$_4$, 0.5 M Na$_2$SO$_4$, 1.0 M NaI, 0.025 M I$_2$		0.71	65.0	0.46	14.0

[a] After Ref. 29.
[b] SC is single crystal, PC is Polycrystalline.

reactions involved, the reader is referred to a number of reviews[23-25] on the subject.

3.3.2. Thin Film Solar Cells

Any photovoltaic scheme which envisages large-scale terrestrial applications must consider the availability of the raw materials. Solar cell technology, therefore, must necessarily be based on thin film form of materials. Thin film devices would typically be about 10–50 μm thick, in contrast to bulk devices which are about 150–250 μm thick. Thin film solar cells are also important because of their low cost of fabrication, large areas, and the possibility of convenient integration with other solar-energy-conversion devices.

Another equally important, but often not equally appreciated, aspect of thin films in relation to photovoltaic technology is their unique growth process which makes it possible to tailor-make materials with desired properties, for example, a band gap in the region of 1.1 to 1.5 eV, a high absorption coefficient, a large minority-carrier diffusion length and a low density of recombination centers (i.e., deep levels and grain boundaries) and matching electron affinities and lattice parameters (in the case of heterojunctions).

Another advantage of thin film solar cells is that it is possible to further develop the tandem-structure approach into a more sophisticated version, the integrated-tandem-solar-cell (ITSC) system, in which the built-in electrical connections provide a permanent series connection between the different cells. Examples of such structures are[26-28]: AlGaAs–GaAs–Ge ($\eta \sim 26\%$), $Ge/Ga_{1-x}In_xAs/Ga_{1-x}In_xP/Cd_{1-x}Zn_xS/ITO$ ($\eta \sim 33\%$), and $CuInSe_2/DdS/AgInS_2$ ($\eta \sim 30\%$).

A factor which limits the performance of thin film solar cells is the polycrystalline nature of thin films in general. The grain boundaries present in polycrystalline films provide recombination surfaces for minority carriers and thus degrade the performance of the device. Further, the grain boundaries affect the device operation by allowing interdiffusion of certain atomic species or diffusion of a particular element from one surface to another, creating shorting paths. Efforts to overcome these problems are the subject of investigation at several laboratories. The electrical activity of the grain boundaries needs to be suppressed, as for example, by passivation (obtained by hydrogenation in the case of Si), by selective anodization of the grain boundaries rendering them ineffective, or by heavy doping with appropriate dopants to provide a direct field away from the grain boundary surfaces. Diffusion through grain boundaries can be minimized or eliminated by providing multilayered diffusion barriers. This concept is utilized in spray-deposited $Cu_2S/CdS–CdS:Al_2O_3$ thin film solar cells where a graded composition multilayer $CdS–CdS:Al_2O_3$ stack prevents the diffusion of Cu from the top Cu_2S layer to the bottom electrode, both during fabrication as well as during subsequent operation, thus lending greater stability to the structure.

A serious problem with thin film devices, particularly $Cu_2S–CdS$ cells, has been the structual, microstructural, and stoichiometry changes during operation, which lead to instability and, finally, degradation. Doping with impurities to stabilize the stoichiometry and structure of the Cu_2S films and thereby arrest degradation has been tried in our laboratory with some success.

Thin film solar cells of several materials have been fabricated in the form of homojunctions, heterojunctions, MIS junctions, and PECs. As shown in Fig. 3.11, for the Cu_2S/CdS combination, thin film solar cells

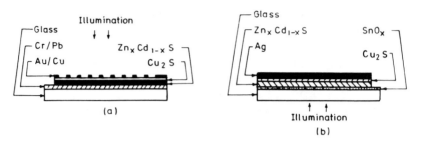

Figure 3.11. Geometrical configurations of a frontwall (a) and a backwall (b) Cu_2S/CdS solar cell.

can be fabricated in the frontwall as well as the backwall geometry. The geometrical design may be chosen so as to protect the unstable layer from exposure to the environment. For a given material, the performance of the device may vary with the deposition technique and the deposition conditions. This is expected since the device performance depends greatly on the microstructure of the films and the topographical features of the interface, which in turn are governed by the mode of deposition and the conditions during growth. Table 3.3 lists the performance characteristics of some of the thin film solar cells (details are given in Chopra and Das[29]).

Table 3.3. Performance of Some Thin Film Solar Cells[a]

Materials	Junction type	Deposition technique	Efficiency (%)
1. Cu_2S/CdS	Hetero	Chem./Evap.	~10
2. $Cu_2S/Zn_{0.16}CdS_{0.84}S$	Hetero	Chem./Evap.	10.2
3. $Cu_{2-x}S/CdS$	Hetero	Evap.	3.3
4. CdSe	MIS	Evap.	5
5. CdS/CdTe	Hetero	CSVT	~10
6. $CuInSe_2/ZnCdS$	Hetero	Evap.	~10
7. Zn_3P_2	Schottky	Evap.	2.5
8. InP/CdS	Hetero	CVD	5.7
9. GaAs	Schottky	MOCVD	5.5
10. GaAs	$n^+/p/p^+$	CLEFT	17
11. Si	Homo	CVD	9
12. Si	Homo	TESS	12
13. a-Si:H	Schottky	Gl.DiS	6
14. a-Si:H:F	MIS	Gl.Dis	6.2
15. a-Si:H	p–i–n	Gl.DiS	10

[a] For details, see Ref. 29.

It should be noted that only recently have $Cu_2S/CdZnS$, $CuInSe_2/CdZnS$, $CdS/CdTe$, and a-Si:H p–i–n solar cells achieved efficiencies of $\sim 10\%$ (on small areas only). However, it should be pointed out that research and development on thin film solar cells have gained impetus only in the last couple of years and since then a steady increase in the conversion efficiencies has been reported. The encouraging upward trend makes us confident that in the next few years thin film photovoltaic technology will advance rapidly and large-area, automated production of devices with module efficiencies exceeding 10% will be a reality. However, it must be emphasized that an unambiguous choice among the existing thin film solar cells and the associated production technologies for large-scale terrestrial applications has yet to emerge.

3.4. Applications in Imaging

The radiation detectors described in Section 3.2. give an electrical output which represents the average incident photon flux. Imaging devices[30,31] are a special class of detectors which provide a visual output representing the spatial variation of the incident photon flux. Any imaging device consists of two main units: (1) the pickup unit, which converts the incident optical (visible or invisible) image into the corresponding electrical image, and (2) the display unit, which converts this electrical image into a visual image. The device is either of direct-viewing or televiewing type, depending on whether these two units are housed together as a single device or separately as two units which may be situated miles apart. A direct-viewing device may operate as an image intensifier or an image converter, or both. In contrast to conventional photography where each part of the object is imaged at the same time, in photoelectronic imaging, except in direct-viewing devices, the object is imaged in small discrete units, which are assembled together in proper sequence in the display unit to form the complete image within a period less than the response time of the eye ($\sim 1/30$ sec).

The heart of a pickup unit is its image sensor and that of a display unit, its display screen. Both photocathodes and photoconductors may be employed as image sensors. The optical image incident on a photocathode is converted by photoemission into a pattern of photocurrents which corresponds exactly to the input radiation pattern. In a direct-viewing device, this pattern of photocurrents is electro-optically projected onto the display screen where it produces a corresponding visual image. The display screen usually consists of a powder coating of a cathodoluminescent material. Although thin film luminescent screens give better resolution and contrast, their efficiencies are lower because of the optical interference effects associated with thin films.

In a television pickup unit (i.e., the camera), since the display unit is situated miles away, the photocurrent image has to be converted into a video signal which can be transmitted. This is accomplished by projecting the pattern of photocurrents onto the insulating surface of a thin film storage target, where the corresponding charge image is stored. The storage target consists of a thin insulator film on a conducting backplate, for example, Al_2O_3 on Al. The charge image is read and picked up by a scanning electron beam in small units. A video signal corresponding to each small element of the charge pattern is generated as a capacitive voltage across a load resistor connected to the back plate (see Fig. 3.12a). Higher photosensitivity and image intensification are obtained[32] by replacing the usual glass target with a thin film (~ 500 Å) of MgO which has a higher secondary emission ratio (~ 7 to 8). This type of storage targets is used in the image-iconoscope type of tubes.

The image-orthicon type of camera tubes use similar targets but without a conducting back plate. Here, the electron-beam scanning is done at the rear surface of the target (instead of the front surface as in the previous case), which discharges the target and produces current variations in the reflected beam which enters a secondary-emission multiplier generating an amplified output signal across the load resistor, as shown in Fig. 3.12b.

Another thin film storage target[33] which is used in very high-gain SEC camera tubes is based on secondary electron conduction (SEC) in porous insulator layers. SEC camera tubes are the most sensitive pickup units capable of operation at very low light levels. An SEC storage target can have secondary emission ratios up to 300. The basic structure of an SEC target is shown in Fig. 3.12c. It consists of an Al_2O_3 supporting layer on an Al back plate (each about 700 Å thick) onto which a very low-density layer of an insulator having good secondary emission characteristics, for example, MgO, KCl, LiF, NaBr, and BaF_2, is deposited. The low-density layer is 20 μm thick with a density only 1–2% of the bulk. Such low-density layers are formed by evaporating the material in an inert atmosphere at a pressure of about 2 Torr.

The vidicon family of camera tubes uses high-resistivity ($>10^{12}$ Ω cm) photoconductors to serve the purpose of both the image sensor and the storage target. High resistivity is very essential for a storage target to prevent lateral flow of charges which causes loss of resolution and image smearing. At the same time, the insulating layer should have sufficient conductivity to allow charges to leak through to the other side, when required, in a time less than the frame time. The photoconductor is deposited on a transparent conducting glass to serve as the back plate. The input optical image projected onto the photocathode through the back

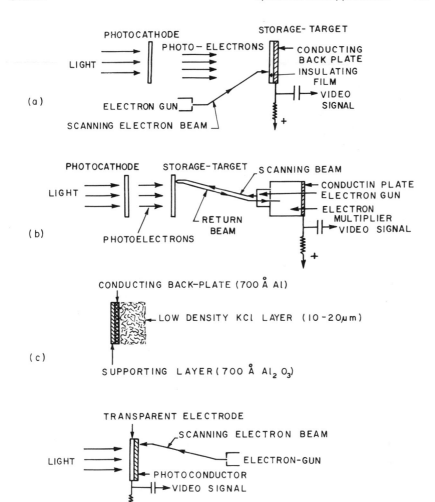

Figure 3.12. Various types of thin film targets: (a) image iconoscope; (b) image orthicon; (c) SEC target; (d) vidicon target.

plate (which is continuously being scanned by an electron beam) produces the corresponding conductivity variations in this layer, causing the previously established surface charges to leak to the back plate which is held at a potential of about + 20 V. Upon recharging of the photoconductor surface to its equilibrium potential by the reading electron beam, output voltage variations proportional to the incident radiation are produced across the load resistor connected to the back plate, as shown in Fig. 3.12d.

Table 3.4. Materials Commonly Used for Vidicon Targets[a] in Different Regions of the Spectrum

Photoconductor material	Region of operation	Photoconductor material	Region of operation
Sb_2S_3	Visible	Ag_2S	Near IR
Amorphous Se	UV, X ray	Au-doped Si	IR
PbO (p-i-n)	X ray visible	PbO–PbS	IR
$Sb_2(S_xSe_{1-x})_3$	*Near IR*		

[a] Vacuum evaporated except for a-Si:H, which is GD prepared.

Besides the requirement of high resistivity to enable charge storage, the photoconductor must have a response time equal to or less than the frame time (1/30 sec) to prevent lag or image smearing in the case of moving scenes. Table 3.4 shows suitable materials for use in vidicons for different regions of the spectrum, along with the method of preparation of the target.

The various photoelectronic devices described above make use of a scanning electron beam for generating the video signal. This means that, in addition to the image sensor, the device will have an electron gun and a number of accelerating and focusing electrodes and deflection coils, thus making the auxiliary unit unduly dominant and bulky in relation to the main sensing unit. Thin film technology has made it possible to fabricate self-scanned image sensors as miniature cameras which may replace the monstrous television camera tubes in the near future. These will be discussed in the next chapter.

3.5. *Electrophotography (Xerography and Electrofax)*

Electrophotography,[34] involving electrostatic reproduction of images, is another application of insulating photoconductor films. The electrophotography plate is similar to the vidicon target described in the previous section. It consists of a thin film of an insulating photoconductor (or a photoconductor in conjunction with an insulator layer) sandwiched between two metal electrodes (Fig. 3.13). The surface of the plate is charged using a corona discharge. The optical image is then projected onto the plate. At the illuminated areas of the plate, the resistivity drops and the charge flows across the photoconductor to neutralize the induced charge at the opposite face of the plate. A latent charge image is thus formed on the plate. The plate is then dusted with black-dyed particles, called the *toner* particles, which fulfill the role of an ink. These particles are held at the unexposed portions by electrostatic forces. Xerography and electrofax processes differ

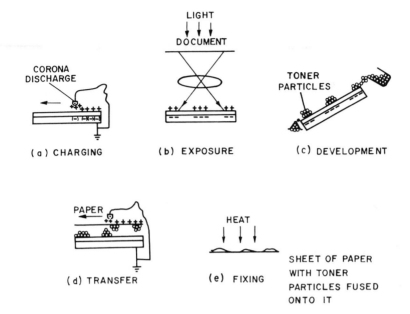

Figure 3.13. Schematic representation of the electrophotographic process.

only in the development of this plate. In electrofax, the photoconductor layer, such as that of ZnO, is deposited directly on a sheet of paper. A fixed image is produced on this paper by heating so as to fuse the toner particles with it. Thus, a photographic film can be used only once. In xerography, on the other hand, the image formed on the photographic plate (consisting of a Se photoconductor film) is transferred onto a separate sheet of paper charged by corona discharge and held over the photographic plate. The toner particles are electrostatically attracted on the paper, where the image is finally fixed by heating. After brushing off any excess toner particles and flooding the plate with light to remove any residual charges, the plate is ready to be used again.

3.6. Thin Film Displays

In photoelectronic displays,[30,35] the image is constructed by electrically addressing each element of the display screen with the corresponding video signal in a total time less than the response time of the eye. In a television display unit, for example, the addressing is done by a scanning electron beam intensity-modulated by the video signal. Although commercialization may take a few more years, the feasibility of self-addressed display screens has been demonstrated using different display elements. These along with the self-scanned image sensors will be discussed in more detail in the next chapter on microelectronic applications. In this section, we describe various types of discrete display elements, and their principle of converting the electrical signal into a visual signal, without going into the details of how an assembly of these elements is addressed to compose the image. Displays are generally classified as emissive and nonemissive. Except for the luminescent type, all others are of nonemissive type.

3.6.1. *Electroluminescent (EL) Displays*

Luminescence[36] is the phenomenon of emission of light from a body (in excess of that from a blackbody at the same temperature) as a result of absorption of energy. Luminescence in a material is associated with the presence of certain specific radiative recombination centers. The luminescent centers may be created either by certain impurity atoms, called the *activator* atoms, in an interstitial or substitutional position of one of the lattice ions, for example, Cu and Ag in II–VI compounds, or by stoichiometric excess of one of the constituents. Light emission occurs when an excited carrier undergoes recombination at these centers. Sometimes it may be necessary to introduce, probably for charge compensation, another type of impurity, called the *co-activator*, for example, Cl, Br, I, Al, or Ga, in II–VI compounds. By varying the amounts of activators and co-activators, ZnS phosphors can be prepared with emission varying from blue to green. When the impurity atom is a transition metal having an incomplete outer shell, local excitation of the electrons in impurity atoms to higher empty levels can take place. De-excitation to the ground state then produces emission characteristic of the transition metal, for example, orange emission in ZnS doped with Mn. Luminescence resulting from the action of an electric field or an electric current through the material is called *electroluminescence*. In electroluminescence, light is emitted by one of the two following processes: (1) field effect or impact-ionization electroluminescence (the Destriau effect[37]) and (2) injection or recombination electroluminescence (the Lossev effect[38]). In the first case, free carriers in the phosphor are accelerated by the externally applied electric field ($\sim 10^6$ V/cm) and acquire sufficient kinetic energy to impact-ionize the luminescent centers. Phosphors based on ZnS, ZnSe, $ZnS_{1-x}Se_x$, $Zn_{1-x}Cd_xS$, and $Zn_{1-x}Cd_xSe$ are some of the common materials exhibiting the Destriau effect. The basic structure of a thin film electroluminescent cell is shown in Fig. 3.14a.

A thin ($\sim 0.5 \mu$m) layer of electroluminescent ZnS:Mn, Cu, Cl is evaporated on a substrate which has been successively coated with a conducting transparent electrode of In_2O_3 or SnO_2 and an insulating layer,

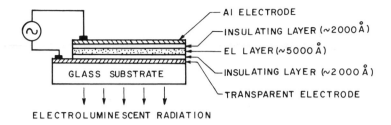

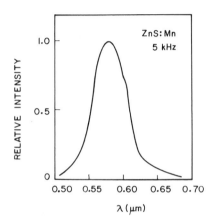

Figure 3.14. Electroluminescent cell (a) and its emission spectrum (b) (after Ref. 39).

for example, of Y_2O_3, Si_3N_4, Al_2O_3, and SiO. Another insulating layer and an Al electrode are then deposited in succession on top of the EL layer. Application of an alternating field between the two electrodes produces luminescence in the half-cycle when the Al contact is negative. Insulating layers are provided to give capacitative coupling of the EL layer to the electrodes. The maximum conversion efficiency of such a cell is ~5%. Figure 3.14b shows the emission spectrum of such a device. The device stabilizes[39] with aging and can give continuous constant illumination for at least 2×10^4 h. The brightness of the device first increases with applied voltage and then saturates. The EL cell can be endowed with an inherent memory due to hysteresis in the brightness vs. voltage curves, which makes it very useful for a large number of applications. EL films, which have a number of prospective applications, are regarded as very promising candidates for the fabrication of flat panel TV screens.

An electroluminescent layer in conjunction with a photoconductor (PC) layer (Fig. 3.15) forms an efficient solid-state image intensifier, giving amplification factors of ~50. Here the PC layer acts as a regulator for EL emission. In dark, the resitivity of the PC layer is much higher than the EL layer so that the field acts across the PC layer. Upon illumination,

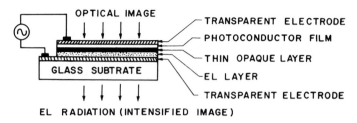

Figure 3.15. Schematic diagram of a solid-state image intensifier.

the resistivity of the PC layer falls, so that a field sufficient to excite luminescence acts across the EL layer. The function of the opaque conducting layer between the PC and EL layer is to prevent optical feedback from the latter to the former. However, if the opaque layer is removed, the device can serve to store the image for some time. Usually the photoconductors used in conjunction with ZnS phosphor are CdS and CdSe. Storage may also be obtained by utilizing[40] the hysteresis behavior in the $I–V$ curves of CdSe-type photoconductors or by using[41] EL layer in conjunction with a ferroelectric layer.

Injection electroluminescence involves injection of excess minority carriers into a semiconductor to which the electrodes are attached directly (in contrast to the Destriau effect where the electrodes are capacitatively coupled to the phosphor) and their recombination with the majority carriers to give energy in the form of visible radiation. Such minority carrier injection can take place at (1) a forward-biased $p–n$ junction and (2) a Schottky barrier (a junction between an n-type semiconductor and a higher-work-function metal, or between a p-type semiconductor and a lower–work-function metal). Application of a forward bias, qV_D, to the Schottky barrier, where V_D is the difference in the work functions of the metal and the semiconductor, allows both minority-carrier injection into the semiconductor (which leads to recombination radiation), as well as majority-carrier extraction, which does not. Majority-carrier extraction can be suppressed and minority-carrier injection enhanced by[42] placing a very thin (50–500 Å) insulating layer of a material with band gap greater than the semiconductor, between the metal and the semiconductor. Injection electroluminescence has been observed in a large number of forward-biased $p–n$ junctions, such as homojunctions are based on SiC, CdTe, GaAs, $Cd_xZn_{1-x}Te$, $Cd_xMg_{1-x}Te$, and $ZnSe_xTe_{1-x}$. Some of the thin film heterojunctions showing injection electroluminescence are $Cu_2S–ZnS$ and $Cu_2Se–ZnSe$. Commercial light-emitting diodes (LED's) are generally $p–n$ junctions based on GaAs, GaP, and other III–V compounds. The layers are prepared by VPE or LPE. LED's are used in a large number of display applications such as alpha-numeric indicators in calculators and watches.

Injection electroluminescence can also be obtained in a reverse-biased *p–n* junction due to avalanche breakdown and subsequent ionization of luminescent centers by impact ionization by hot electrons, but the EL efficiency is much lower compared to a forward-biased junction.

3.6.2. Electrochromic Displays

Electrochromism (EC) is broadly defined as a persistent and reversible color change induced in a material by an applied electric field or electric current. EC devices can be either all solid state or containing a liquid medium. To distinguish them from the all-solid-state EC devices, the latter kind are sometimes referred to as electrochemichromic (ECC). Electrochromism can occur through one of the following mechanisms: (a) color-center creation by electron or hole injection into a defect center, leading to a new absorption band; (b) charge transfer between impurity centers, leading to annihilation of the original absorption band and creation of a new absorption band; (c) the Franz–Keldysh effect, which involves a shift in the absorption edge to longer wavelengths due to field-induced tunneling; and (d) the absorption band shift due to dipole moment change and energy-level splitting (Stark effect). In the ECC category, the various mechanisms responsible for electrochromism are: (e) simple electroredox reaction; (f) electroplating; (g) pH variation; and (h) electroredox reaction involving a solid film. Out of these effects, only (a), (e), (f), and (h) are of practical importance for displays.

One application of color centers created by an electron beam (cathodo-chromism) is the Skiatron cathode ray tube[43,44] in which the target is an evaporated film of KCl. This device has been used in radar and facsimile systems. A preliminary effort was made by Deb[45] to make a photoimaging device using electrochromic WO_3. The device consists of a CdS photoconductor–WO_3 electrochromic double layer, sandwiched between two transparent electrodes. According to Deb, the image is produced by color centers created in the WO_3 film, producing a broad absorption band peaks around $0.9 \mu m$. The image is erased by applying a reverse potential. But since electrochromism in the CdS–WO_3 system is not polarity sensitive, simultaneous recoloration may occur along with bleaching. Deb and Shaw,[46] in a US patent, have suggested that a charge-carrier-permeable insulator layer (for example, SiO, CaF_2, and MgF_2) adjacent to the EC layer renders the EC device polarity sensitive. Another EC device proposed by Robillard[47] for electrophotographic applications consists of a photoconductor–catalyst–semiconductor–catalyst–semiconductor thin film system. The principle of this device is that an image projected onto the PC causes catalytic Ag^+ ions to be injected into a semiconductor oxide (for example, CeO_2, TiO_2, Ta_2O_5, SnO_2) with concentrations corresponding to the radiation pattern of the input image. The catalytic ions then reduce the oxide to a light-absorbing lower state, forming a color image.

The EC solid-state devices discussed above are very slow at room temperatures. Slow response is not a serious problem in ECC devices utilizing liquid electrolytes. Some important ECC devices are described below.

An ECC redox reaction in its simplest form can be written as

$$M^n + e^- + A^+ \rightleftharpoons A^+ M^{n-1}$$
$$\text{(colorless)} \qquad \text{(colored)}$$

where M denotes an EC material containing a metal ion which can exist in different valence states, and A^+ is a mobile cation such as H^+ or an alkali ion. In a simple ECC redox reaction, ions or molecular complexes in a solution are reduced or oxidized to form colored species which are less soluble than the original species so that local precipitation and therefore color persistence is possible. Some important materials exhibiting this effect are: (1) the viologen family (4,4'-dipyridinium) of compounds which become[48] deep purple on reduction, (2) *o*-tolidine (4,4'-diamino-3,3'-dimethylbiphenyl) which become[49] colored on oxidation, and (3) polytungstate anions (PTA) which give[50] a deep blue coloration on reduction and are prepared by dissolving tungsten trioxide, tungstic acid, or tungstates in various acids. In the patent literature, this simplex redox reaction has been reported for use in indicator displays[51,52] and light filters.[53,54]

Electroplating, which involves reduction of metal ions at the cathode to form a metallic coating which can absorb light, can also be used for light valves and display applications. In 1929, Smith[55] described a light valve for use in conjunction with automobile headlamps using KI and KOH solution. A brown coloration which originates at the anode and diffuses throughout the electrolyte is believed to be due to oxidized I_2. Another very interesting display device[56] based on electroplating is an electrically addressable moving indicator for clocks. It employs a $AgNO_3$ electrolyte with a Pt cathode. Silver is plated and deplated successively on an array of electrodes controlled by an electronic switching circuit, thus appearing as a moving indicator in reflection mode.

In the last category of ECC displays, a solid film deposited on an electrode undergoes an electroredox reaction and exhibits intense coloration. A significant system[57,58] is the one utilizing WO_3 films in an acidic solution. When a negative potential is applied to a WO_3-coated electrode in the configuration shown in Fig. 3.16, a deep blue color is created in the WO_3 film due to the reaction

$$WO_3 + (H + e^-)_x \rightleftharpoons H_x WO_3$$
$$\text{(transparent)} \qquad \text{(blue)}$$

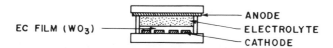

EC FILM (WO₃)

Figure 3.16. An electrochromic display cell.

Recent studies on the ECC mechanism of WO_3 have shown that the coloration is not due to a simple color center creation as proposed by Deb[45] for solid-state WO_3 devices. The ECC WO_3 device described above has been found to withstand several million write–erase cycles and is attractive for application in watch and calculator displays.[59-62]

The EC and ECC displays have quite a few serious problems such as reversibility, side reactions and packaging. There is also a need to develop better electrochromic materials and nonaqueous ECC electrolytes to overcome the problems associated with water electrolysis.

3.7. Information Storage Device

3.7.1. Introduction

This section of the chapter highlights the role of thin film technology in achieving the present state-of-the-art in optically accessed digital computer memories (in short, "optical memories"). Before we describe the various types of optical memories, we would like to familiarize the reader with some of the common terms used in computer technology.

A memory constitutes that part of the computer which receives the data and stores it for retrieval at a later time. In a digital computer, memory consists of a large number of memory cells, each having two stable states and capable of storing a binary digit (or bit) "0" or "1." According to their functional performance, these are of three types: (1) read–write (R/W) memory, (2) read only memory (ROM), and (3) read mostly memory (RMM). Memories, in general, are classified according to the manner in which the data is accessed. In a random access memory (RAM), access can be made to any memory cell in the same period of time because the read (or write) head can see all the cells at the same time, as, for example, when the memory cells are arranged in a planar matrix configuration. In a periodic access memory (PAM), each memory cell appears in front of the detector only periodically, as in disks and drums. In a sequential access memory (SAM), as in tapes, for making access to any memory cell, one has to wait for a certain period of time depending on the position of the memory cell on the tape.

In optical memories, logic "1" is represented as light "present" and "0" as light "absent." Optical memories are of two kinds: (1) those read and written a bit at a time (or the optical hole memories), and (2) holographic memories. As a data storage technology, an optical approach[63] has two important advantages: (1) a high storage density because of the reduction of the size of information to the diffraction limit of light waves, and (2) a high access speed because of the possibility of parallel information processing, as in holographic memories. Another advantage of optical memories is that readout of data is nondestructive (NDRO), resulting in a virtually permanent record in contrast to magnetic media which suffers from signal loss and deterioration due to readout and long-term storage.

3.7.2. Optical Hole Memories

Information can be recorded in thin films of metals like rhodium, or amorphous chalcogenide films on transparent substrates by burning holes[64–66] with a laser beam. Each hole will transmit light when the laser is later used at low powers for readout. Elsewhere, the metal will reflect the light. This kind of memory is of ROM type and is suited for archival and nondynamic storage applications where updating is relatively infrequent, such as microprogram stores, look-up tables, etc. Since reading and writing in this kind of memory generally requires mechanical motion of the storage medium as in the disk memories, the access speed is limited. Hole sizes as low as 0.5 μm have been achieved[67] in some films by laser burning.

3.7.3. Holographic Memories

A holographic memory[68–70] is a nonmechanically addressed, randomly accessible mass memory with archival permanence and electronic accessibility. It consists of an array of small holograms on an unstructured and continuous, erasable medium, each capable of reconstructing a different page of binary data. Illumination of any of these holograms with coherent light generates a real image consisting of an array of dark and bright spots, each representing a bit. This image falls on an array of photodetectors, with one detector element for each bit. Thus, to read a single bit at a particular position in the memory, the laser beam is deflected in a direction to illuminate the appropriate hologram page containing that bit. The output of the proper detected element is then interrogated to determine whether or not a bright spot exists at that location in the image. In a conventional bit-by-bit memory, it is very difficult and costly to achieve the extremely precise registration between the image and the detector, which is very essential to assure that the desired bit falls on its proper location on the detector array. This registration problem is very easily eliminated in holographic memories. Further, the redundancy (i.e., the capability of each point of the hologram to reconstruct the full contents of the page) allows considerable tolerances in the properties of the storage medium. For example, dust particles, speckles, or scratches on the film have very little effect on the image retrieved. No information is really lost until the whole hologram is completely destroyed.

A holographic memory has typically four components: (1) a page composer, (2) a deflector to permit random access to the desired hologram, (3) a storage medium, and (4) a photodetector array to sense the output. These are described in the following sections.

3.7.3.1. Page Composer

A page composer, also known as a digital data block composer, is a device which spatially modulates a light beam in accordance with the signal to be written. It consists of an array of light valves which can be closed or opened electronically. A page composer is essential because it is time consuming to make a new photographic input mask every time a new page is to be recorded. A page composer may be addressed by a modulated scanning-electron beam, by an electroded matrix of conductors on the device, or by CCD shift registers (see Section 4.7) integrated onto the device. Although a large number of choices are available for the fabrication of page composers, only the one based on nematic liquid crystals has shown promising results. Liquid crystals are long-chain molecules having the rheological properties of a liquid but the optical properties of a crystal. In the absence of any field, the molecules are aligned parallel to each other and the material looks transparent. Upon application of an electric field (dc or low-frequency ac) the ordered state is destroyed and the material looks opaque when observed under light, as a result of dynamic scattering. The simplest kind of light valve based on liquid crystals is a layer of liquid crystals sandwiched between transparent conducting glass electrodes. A page composer consists of an array of such valves addressed by an electroded matrix formed by horizontal and vertical conductor strips, as shown in Fig. 3.17a. A particular valve is closed by applying a field between the address conductors.

A light valve can be made[71] light addressed by using the liquid crystal (LC) layer in conjunction with a photoconductor (PC) layer (Fig. 3.17b). This display is ac-driven and is operated in the reflection mode with a multilayer combination of a dielectric mirror and an absorbing layer of CdTe to isolate the CdS–PC from the projection light (or "reading" light). The CdS layer gates the applied ac voltage across the LC film in response to the "writing" light. The page after being composed is recorded on the storage medium holographically. Sometimes the page composer and the detector array are combined together to form an integrated device. An

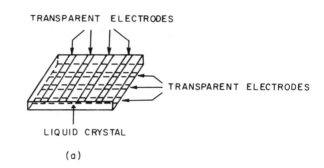

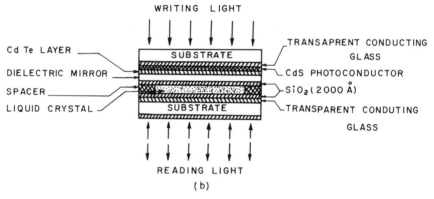

Figure 3.17. Liquid-crystal light valve (a) electronically addressed and (b) light addressed.

advantage of this is that the same reference beam can be used both for reading and for writing of data.

3.7.3.2. Storage Medium

The basic requirement of a storage medium for holographic recording is that the material must be capable of supporting submicron resolution. The various thin film storage devices are described below.

3.7.3.2a. *Magneto-Optic Films.* Holograms can be recorded in ferromagnetic materials by Curie-point writing. Magnetic thin films of MnBi (thickness ∼ 700 Å) approach best the qualities of permanence, indefinite reversibility, high resolution, convenience of writability and erasability by the same beam, and relatively high speed. These films[72–75] are formed in single-crystal form (epitaxial) by vacuum evaporation of Mn and Bi separately onto a mica substrate and subsequent annealing for 72 h in vacuum at 300°C. In operation, the film is first magnetized to saturation in a direction normal to the plane of the film. When the medium is exposed to a holographic interference pattern, at illuminated portions the temperature rises above T_c (Curie temperature = 360°C for MnBi) due to absorption of light, the intensity of which is such that the temperature of the film is always less than the decomposition temperature (450°C) of the film material. On cooling, these regions, in the presence of the demagnetizing fields of the neighboring unheated regions, become saturated in a direction opposite to that of their neighbors. The hologram is now permanently stored in terms of a magnetic pattern until it is deliberately erased as, for example, by the application of a very strong magnetic field in the absence of illumination.

Readout of data is accomplished by directing a reference beam on the hologram. The light transmitted through the film experiences a phase variation corresponding to the magnetization pattern as a result of the Faraday effect. A magnetic hologram therefore behaves as a phase hologram and is reconstructed without the need for polarizers and analyzers. The light reflected from the film also undergoes phase variations as a result of the Kerr effect and similarly yields a reconstructed image.

See also Chapter 5 for more on magneto-optic films.

3.7.3.2b. *Metal Films.* Holograms (amplitude) can also be recorded[76,77] in thin (100–800 Å) metal films of Bi, Sb, Cd, etc. The incident interference pattern is engraved in the film by evaporation of the material due to heating by the laser beam. The hologram can be read out by reflection or transmission.

3.7.3.2c. *Thermoplastic Films.* Thin phase holograms can be recorded[78] in photoconductive thermoplastic films (or a thermoplastic in conjunction with a photoconductor) as surface-relief patterns. A latent charge image is formed when a photoconducting thermoplastic [such as polyvinylcarbazole (PVK) sensitized with Brilliant Green dye] film deposited on a transparent conducting glass whose surface is previously charged is exposed to the optical interference pattern. When this photoplastic is heated to above the softening temperature, the variations in the electrostatic pressure across the photoplastic cause fluid deformation corresponding to the charge image, thus creating a surface-relief pattern which is fixed upon cooling. Erasing of the hologram is done by heating it to higher temperatures. Thermoplastics offer very high resolution, highly sensitive and efficient, reusable and replicable holograms with archival permanence. These holograms are insensitive to ambient light, making handling easier.

3.7.3.2d. *Amorphous Chalcogenide Films.* As already mentioned before, a large number of chalcogenide glasses undergo long-lasting reversible optical changes when illuminated. In some cases, light is absorbed as heat and heat performs the storage, for example, by inducing amorphous-to-crystalline phase transformation. In other cases, light produces free carriers by bond breaking, giving rise to various photoeffects. For example, large physical densification (thickness contraction) effects up to 12% in obliquely deposited $Ge_x Se_{1-x}$, 19% in $Ge_x S_{1-x}$, and 26% in $Ge_{0.25(1-x)} As_{0.25(x)} Se_{0.75}$ films, induced by band-gap illumination, and up to 40% in $Ge_x Se_{1-x}$ by irradiation with 50-keV He^+ ions, have been observed by Chopra *et al.*[79-83] The effect has been explained in terms of strain-induced mechanical collapse of the low-density columnar structure due to the columbic interaction of the photogenerated charge carriers and energetic ions with the intrinsic charged dangling bonds of the chalcogenide network. This conclusion is corroborated by the SEM and TEM studies.

Extensive studies on obliquely deposited S- and Se-based chalcogenide glasses further show that the densification effect is accompanied by changes in the band gap (~ 0.13 eV for photons and ~ 0.20 eV for He^+ ions), refractive index ($\sim 3.5\%$ for photons and $\sim 10\%$ for He^+ ions), rate of chemical dissolution ($\sim 33\%$ for photons and $\sim 88\%$ for He^+ ions), and amount of electrochemically adsorbed metal. A critical interpretation of these effects has revealed that the necessary conditions for the occurrence of large radiation induced effects are: (1) the existence of a well-separated lone-pair band on the top of the valence band, (2) a large density deficit, (3) low bulk density, and (4) a strong electron–phonon coupling. Though the lattice structure and low coordination number of S and Se are conducive to large lattice arrangements, pure films of these chalcogens deposited at 300 K do not show any radiation-induced effects. This is due to the nascent self-bleaching effect because of the low glass transition temperature (298 K for S and 318 K for Se). However, if deposited and irradiated at low temperatures, pure S and Se films do exhibit large radiation-induced effects.

The large radiation effects in obliquely deposited chalcogenide films can be utilized for both positive and negative imaging in reprographic and submicron lithographic applications in the following four modes: (1) photo-doping, in which a chalcogenide film with electrochemically adsorbed silver is irradiated through a desired mask and developed by wet or dry etching, to yield a negative image; (2) photochemical process, in which the 0°-deposited film, after exposure through a desired mask and development, yields a positive image and the 80°-deposited film yields a negative image; (3) electrochemical doping, in which silver is electrochemically adsorbed on irradiated 80°–deposited GeSe film which is then developed to yield a negative image; and (4) thickness contraction, in which the irradiated region contracts in thickness to give a positive relief pattern. The relief pattern may be further developed due to the selectivity of etching (dry or wet) of the irradiated region.

The photodoping mode yields a resist of sensitivity ~ 1 J/cm^2 (photons) and $\sim 10^{-5}$ C/cm^2 (ions) with contrast ~ 4 (photons) and ~ 8.3 (ions) and with a resolution ~ 1 μm. A bilayer system utilizing the high sensitivity of the organic photoresist as a top layer for patterning has been used to generate $\sim 1.5\ \mu m$ linewidths with $\sim 0.2\ \mu m$ edge resolution on SeGe films.

Photoeffects which have been utilized for storage are: (1) amorphous-to-crystalline phase transformation, (2) refractive index changes, and (3) photodoping. Crystallization can be due to thermal effect or photoeffect of radiation. Spot storage utilizing amorphous-to-crystalline phase transition has been demonstrated[84] in sputtered thin films of $Te_{81}Ge_{15}Sb_2S_2$ with an argon laser.

Hologram storage has been accomplished by several investigators[85–87] in thin films of As_2S_3 and related materials. The effect utilized for recording in these materials is the refractive-index change upon illumination. These volume holograms can be converted to surface-relief holograms[88] by etching in a basic solution. Overcoating of a suitable metal (which does not diffuse thermally into the chalcogenide film), such as Ag and Cu, gives these holograms a new degree of permanence. These metals are known[89] to diffuse into films upon illumination.

It may be mentioned here for the sake of completion, that holographic storage has also been demonstrated in photoconductor–ferroelectric single-crystal (bismuth titanate)[90,91] and in photoconductor–electro-optic single crystal (PLZT)[92,93] structures. Storage may also be obtained[94,95] in photo-conductor–liquid crystal (90% nematic and 10% cholestric) composite structures. In principle, it should be possible to obtain similar results in suitably deposited films of these materials.

3.8. *Amorphous Silicon-Based Devices*[96]

Glow discharge decomposition of silane yields hydrogenated amorphous silicon (a-Si:H) films. By controlling deposition parameters, such films with a dark resistivity as high as $10^{13}\ \Omega$ cm, which decreases to $10^6\ \Omega$ cm in light, have been prepared. These characteristics are well-suited for electrophotography applications. Using a multilayered structure of the type glass/SnO_2:F/a-Si:H/Sb_2S_3, video cameras and vidicon tubes of excellent performance characteristics have been demonstrated.

4

Microelectronic Applications

4.1. *Introduction*

The need for miniaturized electronic components in the microelectronics industry has provided the greatest stimulus for investigation of the electrical properties of thin films of various materials. Be it the fabrication of monolithic semiconductor integrated circuits or thin film integrated circuits, the role played by thin film technology is indispensable. In the conventional technology of integrated circuits (ICs), both the active and the passive components are formed into an active substrate of single-crystal Si by the conventional diffusion and thermal oxidation techniques. Electrical contacts and interconnections to these components are provided by thin film deposition techniques. In thin film IC technology, all the components as well as the interconnections are deposited by thin film techniques on a glass or ceramic substrate. Complicated integrated circuits employing millions of thin film components can be fabricated in a single pumpdown cycle of a vacuum system using appropriate masks. Thin film components such as resistors and capacitors can also be used as discrete devices with much superior performance as compared to their bulk counterparts. In the present chapter, we shall first describe the various active and passive thin film components and then discuss some examples of complete thin film ICs. Various other microelectronic devices are discussed individually in the remaining sections.

4.2. *Thin Film Passive Components*

Before we proceed any further with devices, let us briefly review the electronic properties of thin films which are relevant to device fabrication. This discussion is divided into two sections, viz. metal films and insulator films. Results are given only in the form of final expressions since the detailed treatments are given in a number of textbooks.[1–5]

4.2.1. *Electrical Behavior of Metal Films*

The electrical resistivity of metals, which in bulk form arises from the scattering of conduction electrons by lattice vibrations (phonon scattering), impurities, and structural imperfections if present, is greatly increased in thin films. The increase is due to (1) the surface scattering of conduction electrons at the film surfaces (thickness or size effect), and (2) the presence of a large number of impurities and imperfections inherently introduced during the film growth process. The size effect due to surface scattering arises when the thickness of the film is of the order of or less than the mean free path of the electrons. The expressions for the electrical resistivity ρ_F and the temperature coefficient of resistance (TCR) α_F for a thin film, as derived by Fuchs[6] and Savornin,[7] are given by

$$\rho_F/\rho_B = \begin{cases} 1 + (3/8k)(1 - p), & \text{for } k \gg 1 \text{ and } \forall p \\ (3k/4)(1 + 2p) \ln (1/k), & \text{for } k \ll 1 \end{cases} \tag{4.1}$$

and

$$\alpha_F/\alpha_B = \begin{cases} 1 - (3/8k)(1 - p), & \text{for } k \gg 1 \\ 1/\ln (1/k), & \text{for } k \ll 1 \end{cases} \tag{4.2}$$

where the subscript B refers to bulk, k is the ratio of the film thickness to the electron mean free path, and p is the fraction of electrons reflected specularly at the film surface. It should be noted that the TCR values [defined as $\alpha = (1/R)(dR/dT)$, where R is the resistance at temperature T] refer to temperatures above the Debye temperature where resistivity varies linearly with temperature. Below the Debye temperature, the resistivity varies as T^5 for bulk metals and as T^3 for thin films. From Eqs. (4.1) and (4.2), it can be deduced that the size effect in thin films (which obviously vanishes if all the electrons are specularly reflected, that is, $p = 1$, which may occur when the surface roughness of the film is less than the de Broglie wavelength of conduction electrons) causes the resistivity to increase and TCR to decrease. Even though this result seems very attractive for fabrication of thin film resistors, the phenomenon of size effect is rarely used in practice because it occurs only in ultrathin films which pose stability and control problems.

A much greater contribution to resistivity is made by grain-boundary scattering,[8] which occurs when the grain size of films is of the order of or less than the electron mean free path. The presence of impurities and imperfections in concentrations greatly in excess of the thermodynamic equilibrium values also leads to high resistivity values with low TCR. Such films, however, tend to exhibit long-term drift in resistivity due to self-annealing-out of defects and require suitable heat treatments for stabiliz-ation. Depending on the conditions under which the heat treatment is given, its effect may be (1) to reduce the resistivity due to decrease in the concentration of defects and imperfections and/or grain growth during annealing, or (2) to increase the resistivity due to oxidation effects and agglomeration (present in ultrathin films only).

Similar to agglomerated films, island-structured discontinuous ultra-thin films formed in the initial stages of growth also have much higher resistivities than the continuous ones. The low electrical conductivity of such films is generally characterized by a negative TCR arising as a result of a balance between the positive TCR associated with the metal islands and the negative TCR with the electron transfer between the islands. A positive TCR, however, may be observed under special conditions where the coefficient of thermal expansion of the substrate is much higher (for example, Teflon) than that of the metal film. The conductivity varies exponentially with the inverse of temperature, suggesting that the conduc-tion mechanism is thermally activated. It exhibits an ohmic behavior at low values of applied field E and a nonlinear behavior following an $\exp (E)^{1/2}$ dependence at high fields. The various conduction mechanisms proposed to explain the observed electrical behavior of discontinuous thin films, such as thermionic (Schottky) emission and thermionic emission accompanied by tunneling through the gap between the islands, either directly or via traps in the underlying dielectric substrate have been reviewed by a number of workers.[9–10] Discontinuous films are, however, very susceptible to oxidation and tend to agglomerate. Successful resistors using discontinuous thin films have been reported only for the case of rhenium. Because of their high strain sensitivity and large TCR values, discontinuous thin films can be used for fabrication of strain gauges and temperature sensors. Another phenomenon which can give rise to negative TCR values in thin metal films is grain-boundary oxidation. This is so because such films, even though physically continuous, are electrically discontinuous due to intergranular insulating barriers.

Yet another possible source of high resistivity in thin films is their porous or network structure obtained under special deposition conditions. These films are electrically similar to two-phase metal–insulator systems (cermets) where the conducting metal is "diluted" by dispersing it in an insulator matrix, so that the electrical conduction is controlled by the interconnecting regions. Porous films are again very susceptible to oxidation because of their large surface area. However, with suitable protection these films may have very high resistivity with very low TCR and adequate stability.

In addition to their use as resistors, and conductors for interconnections and contacts, metal films can also be used for fabrication of inductors, as will be discussed later.

4.2.2. Dielectric Behavior of Insulator Films

In contrast to resistivity, where an intrinsic size effect arises due to surface scattering of conduction electrons, there is no such effect[11] in the case of dielectric properties of a material. This means that bulk values of dielectric constant and dielectric loss should be retained in thin films, provided, of course, the films are structurally perfect. In the ultrathin range, however, because of the presence of discontinuities, pinholes, and discrete defects, insulating films start exhibiting marked differences from the bulk. For example, porosity in thin films leads to a rapid fall in the dielectric constant. An apparent thickness dependence of the dielectric constant in ultrathin films ($\leqslant 50$ Å) may also be observed because of a reduction in the effective thickness of the insulator layer due, for example, to penetration by energetic metal-vapor atoms during deposition of electrodes for measurements. The dielectric loss, which occurs when electric polarization (be it electronic, ionic, or dipole) in an insulator is unable to follow the electric field variations, is again expected to be independent of film thickness unless the defect structure takes part in a particular relaxation process or the film is thin enough to influence the internal reorientation of dipoles. Diffusion of electrodes into the insulating layer may also produce a significant increase in the measured values of the dielectric loss.

Because of the geometry of thin films, a number of intrinsic mechanisms of dielectric breakdown present in bulk insulators are precluded in thin films, making the breakdown phenomenon much simpler. This leads to breakdown fields $\sim 10^7$ V/cm in the best insulating films compared to 10^6 V/cm in the best bulk insulators (except for mica and a few plastics). In systems where avalanche breakdown (i.e., breakdown due to ionization of the lattice by collisions with energetic electrons) is the dominant breakdown mechanism, an intrinsic size effect resulting in increase in breakdown field may arise in films with thickness of the order of or less than the mean free path, due to the increase in the number of inelastic collisions with the lattice. This type of size effect has been observed[12,13] in ZnS and Al_2O_3 films. However, in practical devices such as capacitors, where much thicker films (compared to the mean free path) are used, a number of extrinsic breakdown mechanisms due to structural imperfections can be present and are the major cause of breakdown and device failure. These failure mechanisms along with the methods of combating them will be discussed in Section 4.2.4.

4.2.3. Resistors

A thin film resistor consists of a resistive path deposited between two contact electrodes on an insulator substrate, generally in the configuration

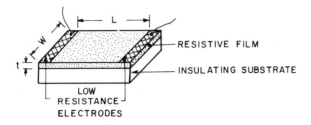

Figure 4.1. Common geometry of a thin film resistor.

shown in Fig. 4.1. The resistance R of the resistor in this geometry is given by

$$R = (\rho/t)(L/W) \qquad (4.3)$$

where ρ is the specific resistivity of the material, L and W are, respectively, the length and width of the resistive path, and t is the thickness of the film. For comparing different resistors, a normalized measure of resistance, called the sheet resistivity R_s (ohms/square), which depends only on the thickness of the film for a given material, is used. The quantity R_s is given by

$$R_s(\Omega/\square) = \rho/t \qquad (4.4)$$

The actual resistance value R of the resistor in terms of R_s is therefore given by

$$R = R_s N \qquad (4.5)$$

where N is the number of squares of area W^2 ($W < L$).

In any resistor technology in which precision components are desired, one of the major steps in the fabrication of resistors is trimming to the required design value. Equations (4.3) and (4.4) show that this can be done either by adjusting the dimensions of the film or by changing its resistivity. The resistance value is generally kept below the design value and later trimmed to the required value. In cases where sheet resistance is constant over the entire substrate, all the resistors on it can be trimmed in one step (substrate trimming) either by (1) a heat treatment, which changes the resistivity through annealing or through grain boundary oxidation; or (2) thickness reduction by RF sputter etching or anodic oxidation (wherever possible). The latter has an additional advantage of providing a protective layer on top of the resistor. In cases where substrate trimming is not feasible, as when local sheet resistance varies from one resistor to other, individual resistor trimming is employed. In addition to heat treatment and anodic oxidation, an individual resistor may also be trimmed by changing its geometry by removing a portion of the film completely as, for example, by burning using a laser beam (laser trimming) or by abrasion. Whereas trimming by heat treatment also changes the TCR, trimming by other means does not.

The resistor requirements for most applications are met to a large extent by films with sheet resistance in the range 10–1000 $\Omega/\square$, corresponding to a resistivity range of 100–2000 $\mu\Omega$ cm. In addition to a proper resistivity value, the material must also possess a low TCR (< 100 ppm/°C) and should have high thermal, mechanical, and aging stability and easy control of the value of resistivity. Although the resistivity of metals (except for transition elements) and their alloys in bulk form rarely exceeds 20–30 $\mu\Omega$ cm, it can be significantly increased in thin film form, as has been discussed in Section 4.2.1. Semiconductors, on the other hand, do have the required resistivity values even in the bulk form, but the large negative TCR values associated with them makes them unattractive for resistor applications.

A large number of metals have been investigated for their use in resistors. Of these the ones of great importance are the refractory metals. Exceptionally stable resistors can be made from refractory metals by sputtering them in controlled atmospheres. Their refractory nature provides thermal stability to films, implying that the imperfections frozen in during deposition do not anneal out during their operational life. Stability against oxidation is provided due to their ability to form tough, self-protective oxides on the surface either by thermal oxidation or by anodization. Amorphous-like/disordered thin films of W, Mo, and Re deposited by ion-beam sputtering, having[14] a resistivity ~ 100 $\mu\Omega$ cm and a very small (< 50 ppm/°C) TCR, are attractive for resistors. However, of all the refractory metals, tantalum has attracted the most attention because of the compatibility of these resistors with thin film capacitors and insulated crossovers using tantalum oxide, making it possible to base an RC technology completely on Ta.

One form of Ta having much higher resistivity than the bulk is[15] low-density porous Ta, sputtered in Ar under special conditions, but this has not been found useful for resistor applications because of the associated stability and reproducibility problems. Much better results are obtained by sputtering Ta in a controlled atmosphere of Ar–O_2 and Ar–N_2 mixtures. The resistivity and TCR of these films is strongly dependent on the partial pressure of O_2 and N_2. Figure 4.2 illustrates[16] the effect of partial pressure of N_2 on the resistivity and TCR of Ta films. It can be seen from the figure that a resistivity of ~ 250 $\mu\Omega$ cm with a TCR of about -75 ppm/°C can be obtained under suitable conditions (compared to bulk values of about 13 $\mu\Omega$ cm and 3800 ppm/°C). Resistors in the 10–100 $\Omega/\square$ range can be fabricated using these films. It has also been found that the film composition corresponding to Ta_2N exhibits maximum stability during load-life tests. These tantalum nitride films can be anodized as easily as Ta. Although these films do not self-anneal, their high susceptibility to oxygen requires a stabilization treatment, which involves anodization followed by about 5 h of heat treatment at 250–400°C. Diffusion of oxygen into the grain boundaries during heat treatment, which makes the resistivity unstable and increases capacitive losses at high frequencies, can be prevented[17] by a prior diffusion of Au into the Ta grain boundaries. Ta films sputtered in $Ar-O_2$ mixtures also have similar electrical characteristics but are not as stable and reproducible as Ta_2N films.

Rhenium is another refractory material that has been used for fabrication of resistors. These differ from Ta resistors in that the films used are ultrathin and the high resistivity arises due to discontinuities in the island-like structure of the films. The films are deposited by evaporation onto glass, ceramic, or silica substrates at 500°C in a vacuum of $\sim 10^{-5}$ Torr, and are characterized[18] by a sheet resistance of 250–300 $\Omega/\square$ with zero TCR and 10,000 $\Omega/\square$ with -400 ppm/°C. Good stability is achieved with a protective overlayer of 2000 Å thick SiO.

The second class of materials suitable for resistors are[19] the alloys of the type 80% Ni + 20% Cr (nichrome) ($\rho \sim 110 \, \mu\Omega$ cm, TCR $\sim +85$ ppm/°C), 75% Ni + 20% Cr + 3% Al + 2% Cu ($\rho \sim 130 \, \mu\Omega$ cm,

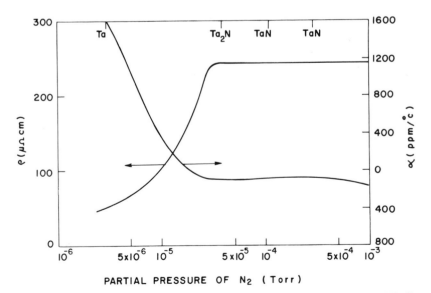

Figure 4.2. Effect of partial pressure of N_2 on the resistivity and TCR of sputtered Ta films (after Ref. 16).

TCR $\sim \pm 20\,ppm/°C$) and 76% Ni + 20% Cr + 2% Al + 2% Fe ($\rho \sim$ 133 $\mu\Omega$ cm, TCR $\sim \pm 5\,ppm/°C$). These have high resistivity and low TCR because of the presence of a transition metal in which the partially filled d-band overlaps the s-band. The density of states in the d-band is higher than that in the s-band at the Fermi level, so that a high probability exists for the conduction electrons to be scattered from the s-band into the d-band where they no longer contribute to conduction. With increase in temperature, the number of electrons scattering into the d-band decreases, thus lowering the resistivity. Over a narrow range of temperatures, this decrease in resistivity compensates the increase due to phonon scattering, resulting in low TCR. The most widely used alloy in industry is nichrome, thin films of which can be deposited by vacuum evaporation[20,21] or sputtering.[22] Thin film resistors using electroless nickel, which is an alloy of Ni and P (0.5 to 13%) have also been reported.[23] These films can have sheet resistance up to 2000 $\Omega/\square$ with TCR less than 100 ppm/°C.

Cermets are another important class of materials for resistor fabrication. A large number of such metal–dielectric systems have been investigated, the most promising being Cr–SiO. Thin films of these materials can be prepared by flash evaporation,[24] RF sputtering,[25] and diode sputtering.[26] The resistivity of these films is strongly dependent on the composition of the cermet and the amount of disproportionation and lies anywhere between 10^3 to 10^5 $\mu\Omega$ cm. An important property of cermets is that high resistivities are obtained without large negative TCR values. For example, a 50:50 mixture of Cr–SiO has a resistivity of 10^3 $\mu\Omega$ cm with a positive TCR very close to zero. When heated in air, unprotected Cr–SiO films are subject to oxidation and undergo increase in resistance. Even though excellent stability can be obtained in unprotected Cr–SiO films if given a 1-h heat treatment at 400°C, a protective layer such as that of SiO or sputtered SiO_2 is generally preferred.[27]

In cases where large negative TCR values can be tolerated, semiconductors can also be used for resistor fabrication. The most successfully used semiconductors are carbon and SnO_2. Because of the high temperatures ($\sim 1000°C$) involved during deposition[28] of carbon films prepared by pyrolysis of C-bearing gases, such as methane, these films are used only for fabrication of discrete resistors and not in microelectronic ICs. Pure carbon films deposited on ceramic substrates have a TCR of the order of -250 ppm/°C at 10 $\Omega/\square$ and -400 ppm/°C at 1000 $\Omega/\square$. By including a boron-containing gas along with CH_4, TCR values are lowered to -20 ppm/°C at 10 $\Omega/\square$ and to -250 ppm/°C at 1000 $\Omega/\square$.

SnO_2 is a very important material in that it is refractory and has no oxidation problems. Extremely adherent, chemically resistant films of SnO_2 are prepared by the method of spray pyrolysis[29,30] of an aqueous solution of $SnCl_4$. The undoped films have high resistivity ($R_s \sim$ few thousand $\Omega/\square$

Table 4.1. Characteristics of Various Materials of Interest for Resistor Applications

Film material	Deposition technique	Sheet resistance $R_s(\Omega/\square)$	TCR ppm °C
Cr–Ni (20:80)	Evaporation	10–400	100–200
Cr–Si (24:76)	Flash evaporation	100–4000	±200
Cr–Ti (35:65)	Flash evaporation	250–600	±150
Cr–SiO (70:30)	Flash evaporation	Up to 600	−50 to −200
SnO_2	Spray pyrolysis/CVD	Up to 10^4	100
W, Mo, Re	Sputtering	10–500	−20 to −100
Ta	Sputtering	Up to 100	±100
Ta_2N	Reactive sputtering	10–100	−85

for films as thick as 1 μm) and TCR $\sim$ ± 200–500 ppm/°C. By doping with Sb or F, the sheet resistivity can be lowered to as low as 5 $\Omega/\square$ with a corresponding TCR of $\sim$ 100 ppm/°C. Table 4.1 summarizes various materials which are important for thin film resistors.

4.2.4. Capacitors

A thin film capacitor is generally a three-layer system consisting of an insulating layer sandwiched between two metal electrodes. The capacitance C of such a parallel-plate condenser of surface area A (i.e., the area over which the top and bottom electrodes overlap) is given by

$$C = \varepsilon_0 \varepsilon A / t \tag{4.6}$$

where ε_0 is the permittivity of free space ($\varepsilon_0 = 8.85 \times 10^{-12}$ F/m), ε is the dielectric constant of the insulator, and t is its thickness. In comparing the properties of different capacitors, it is often convenient to define another term $\bar{C}$, the capacitance density, as

$$\bar{C} = C/A = \varepsilon_0 \varepsilon / t \tag{4.7}$$

Obviously a maximum value of $\bar{C}$ is desirable for miniaturization. In addition to capacitance, another important characteristic of a capacitor is the temperature coefficient of capacitance γ_c (TCC) defined as

$$\gamma_c = (1/C)(\partial C/\partial T)$$

for a thermally isotropic material. The quantity γ_c is related to the temperature coefficient of permittivity $[\gamma_p = (1/\varepsilon)(\partial \varepsilon/\partial T)]$ and the linear thermal expansion coefficient γ_l by the relation

$$\gamma_c = \gamma_p + \gamma_l$$

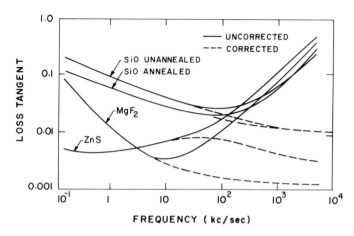

FREQUENCY (kc/sec)

Figure 4.3. Loss tangent vs. frequency for 1000-pF evaporated capacitors of SiO, MgF$_2$, and ZnS. The solid lines are the measured values and the dotted lines are corrected for electrode resistance (after Ref. 31).

In capacitor applications γ_c should be as small as possible. The third characteristic of a capacitor is the dissipation factor, which for thin film capacitors shows[31] a general behavior of increasing loss at very low and very high frequencies (as shown in Fig. 4.3), with a wide minimum in the intermediate frequency range, as a result of superposition of various loss mechanisms. Losses in a bulk capacitor arise due to: (1) the dielectric not being a perfect insulator, that is, the losses due to the leakage resistance R_p of the capacitor, which is placed in parallel with C (this dc contribution to the dissipation factor is given by $1/\omega R_p C$, where ω is the angular frequency and is dominant at low frequencies); (2) ac dielectric loss consisting of a frequency-independent contribution (at room temperatures) present at all levels of loss, superimposed by ac relaxation peaks. In those cases where the majority of the current is carried by defects, as is true in typical dielectric films, the frequency-independent loss is thought to be due to a wide range of distribution of relaxation processes in the film. However, in ideal materials, such as single crystals and amorphous films, it has been suggested that the most likely mechanism is impurity conduction and ionic conduction. The ac relaxation peaks, which generally occur below 1 Hz and very rarely between 100 Hz and 1 MHz, can be due to a variety of causes such as interfacial polarization effects consisting of relaxation of charge buildup at the metal–dielectric interface, or the structural vibrations of the lattice itself. As the temperature is raised, these peaks shift to higher frequencies. Besides the leakage losses and the dielectric losses, there is yet another source of loss which is very dominant in thin film capacitors.

This arises from the resistance R_s of the electrodes, which is placed in series with the capacitance C. The loss due to electrode resistance is given by $\omega C R_s$ and is significant at high frequencies. Within certain limits, the high-frequency characteristics of thin film capacitors can be improved by using thicker and highly conductive electrodes.

A high yield of stable thin film capacitors with high breakdown fields requires continuous, mechanically stable and homogeneous dielectric films and suitable electrodes. Discontinuous areas, pinholes, voids, defects, etc. formed in the film due to the presence of dust particles on the substrate prior to deposition, spitting of the evaporant, or due to some local inhomogeneities act as weak points in the capacitor and give rise to the onset of breakdown. The breakdown characteristics and hence the yield are also influenced by the substrate surface and the electrode material. Metals with a high vaporization temperature give high percentages of failure, probably due to penetration of the dielectric by the energetic incoming metal atoms. It has been established[5] that the best yields are obtained on smooth glass or glazed ceramic substrates with Al electrodes. When thin electrodes (~ 1000 Å) are used, Al has an additional advantage of providing a mechanism for self-healing of weak points in the film. The weak points are isolated by the local melting of the Al electrodes at these points due to excessive current, thus allowing the remaining film to operate satisfactorily.

A large number of dielectric materials have been studied from the point of view of their application in capacitors. Table 4.2 shows some of the important ones along with their dielectric properties. The most successful materials used in thin film capacitors are vacuum-evaporated SiO and anodic Ta_2O_5, both being amorphous. The properties of SiO films are very sensitive to deposition conditions. Best results (ε ~ 4.8 to 6.8, TCC ~ 100 to 400 ppm/°C, dissipation factor in the audiofrequency range from 0.01 to 0.1, and dielectric strength ~ 1 to 3×10^6 V/cm) are obtained[32] for SiO films deposited at high rates, low pressures, and source temperatures around 1300°C. Capacitance densities ranging from 0.018 to 0.0018 μF/cm^2 can be obtained for thicknesses from 0.3 to 3 μm. These values of $\bar{C}$ are relatively small and are useful in high-frequency microwave circuits requiring capacitors of 10 to 1000 pF. Leakage currents as low as 10^{-8} A for capacitor areas of 0.3 cm^2 at 10 V can be obtained. The capacitor yield is also increased by depositing a 2-μm-thick layer of SiO below the lower electrode, which serves to cover the blemishes of the underlying substrate. SiO_2 dielectric layers have applicability in Si technology for capacitors, insulated conductor crossovers, and as diffusion masks for surface-passivation applications. Discrete SiO_2 capacitors are made using either reactively sputtered SiO_2 or RF-sputtered SiO_2 (from a fused quartz cathode) on glass substrates. SiO_2 films are typically characterized[33] by

Table 4.2. Dielectric Properties of Thin Insulating Films of Various Materials for Capacitor Applications

Film material	Dielectric constant	Dissipation factor	Frequency (KHz)	Breakdown voltage (V/cm)	TCC (ppm/°C)	Thickness (μm)	Deposition technique[b]
Al_2O_3	9	0.008	1.0		300	1.5	Ev
$BaSrTiO_3$	10^3–10^4		1.0				Flash Ev
SiO	~3–5	0.004–0.02	0.1–1000	3×10^5	300	~2	Ev
	6	0.015–0.02	1–1000				Ev
SiO_2	3–4	0.001		~10^6			RS
Si_3N_4	5.5			10^7		0.03–0.3	CVD
Ta_2O_5	25	<0.01	0.1–50	6×10^6	250		An
TiO_2	50	0.01	1.0		300		An
W_2O_5	40	0.6					An
CaF_2	~3	0.05	0.1				Ev
LiF	~5	0.03	0.1				Ev
Nb_2O_5	39	0.07	1.0				RS
Polystyrene	~2.5	0.001–0.002	0.1–100				GDP
Polybutadiene	~2.5	0.002–0.01	0.1–100				UVP
$BaTiO_3$	~200	0.05	1.0				Ev

[a] If not mentioned, the thickness is ~1 μm. Al (or Au) films are generally used as electrodes for measurements (after Ref. 1).

[b] Ev, evaporation; RS, reactive sputtering; CVD, chemical vapor deposition; An, anodization; GDP, glow-discharge polymerization; UVP, ultraviolet polymerization.

$\varepsilon \sim 3.9$, dissipation factor ~ 0.0003 (at 1 KHz), TCC ~ 10 ppm/°C (between -200 and 100°C), and breakdown fields $\sim 6 \times 10^6$ to 10^7 V/cm.

Anodic Ta_2O_5 is another material which has been successfully exploited[34] for fabrication of thin film capacitors. Anodic Ta_2O_5 films are amorphous, pore-free, and chemically resistant; they are characterized by $\varepsilon \sim 26$, dissipation factor between 0.002 and 0.01 for frequencies lying between 0.1 to 10 KHz, and TCC ~ 200 ppm/°C between 0 and 100°C. The breakdown field is $\sim 4 \times 10^6$ V/cm if the top electrode is Au, and $\sim 10^6$ V/cm if it is Al. A capacitance density $\sim 0.1 \ \mu F/cm^2$ for a dielectric thickness of 2440 Å can be obtained using anodic Ta_2O_5 films. The main losses in a Ta_2O_5 capacitor arise from the resistance of the bottom Ta electrode. These capacitors exhibit asymmetry in the conduction characteristics. Addition of about 1 at.% Mo to Ta_2O_5 films by cosputtering makes the conduction characteristics almost nonpolar.

The quality of Ta_2O_5 capacitors (Fig. 4.4a) can be improved by inserting a 2000–5000 Å thick semiconducting layer of manganese oxide (generally a mixture of insulating Mn_2O_3 and conducting β-MnO_2, with resistivity between 5 and 10^4 Ω cm), prepared by pyrolysis[34] of $Mn(NO_3)_2$ solution at 200°C or by reactive sputtering,[35] between Ta_2O_5 and the top electrode, as shown in Fig. 4.4b. Defects in the Ta_2O_5 film are self-annealed by the presence of manganese oxide, which acts as a solid electrolyte for anodization of defective areas in Ta_2O_5. Further, the heat generated during local breakdown in the Ta_2O_5 layer can cause reduction of the conducting β-MnO_2 to insulating Mn_2O_3 which protects the defective areas, thus

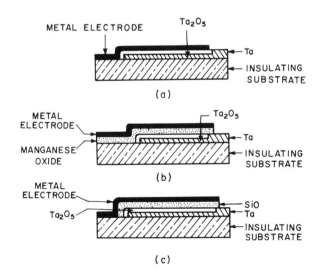

Figure 4.4. Different types of Ta thin film capacitors.

increasing the yield of short-free capacitors. Use of manganese oxide also allows the use of thinner Ta_2O_5 films to obtain higher capacitance densities and improves the high-temperature stability of capacitors. Thus capacitors with capacitance up to several microfarads can be obtained with good yields. However, capacitors using manganese oxide layers have somewhat higher values of dissipation factor and TCC.

For obtaining low-value capacitors in tantalum IC technology, a duplex dielectric structure (Fig. 4.4c) of the type Ta–Ta_2O_5–SiO–Au (or NiCr/Au) corresponding to two capacitors in series has been described[37] in the literature. An important advantage of this duplex structure is a higher dielectric strength than that of each dielectric alone because it is unlikely that the defects in the two layers are at the same sites.

Thick and thin films of a number of organic dielectrics (see Table 4.2) have also been found suitable for capacitor applications. These have the advantage of good mechanical flexibility, high breakdown voltages, and low dissipation factors.

The shape of the electrodes in thin film capacitors is generally rectangular. However, subsidiary electrode structures may be incorporated in the top electrode to provide trimming facilities. Use of interdigited electrode structures has also been suggested for improving the high-frequency response of electrodes.

4.2.5. Inductors

A thin film inductor consists of a spiral track of high conductivity on an insulating substrate. An insulated crossover is used to make the inner connection. The material requirements for inductors are essentially the same as for conductors (see next section) because high resistance values give rise to low quality factors (Q). Thin film inductors, except for values of the order of few microhenries, have not met with much success because of the problem of the physical dimensions of the spiral (which for a 3.5-μH inductor with a Q of about 50 at 15 MHz has a diameter of 7.6 cm) exceeding practical limits. Thin film inductors can therefore be used only in high-frequency circuits requiring low-value inductors. Miniature wire-wound inductors are preferred otherwise.

4.2.6. Conductors (Interconnections and Contacts)

In microelectronic circuits, strips of thin conductor films are used as wiring for the interconnection of various components deposited on the substrate. The basic requirements on the material for interconnections are[38] obviously low resistivity ($\leq 4\ \mu\Omega$ cm), good substrate adhesion, and chemical, mechanical, and thermal stability. In an electronic circuit, it is also required to make external connections to leads or any other discrete component (if necessary) either by soldering or microwelding techniques. Therefore the material used for interconnections should preferably be solderable so that the same can also be used for contacts, thus simplifying the fabrication process. The material used for interconnections must also be capable of being etched to line-widths of the order of a few microns using conventional lithographic pattern generation techniques (see Chapter 1). Lastly, the metal used must be capable of carrying current densities $\sim 3 \times 10^5$ A/cm^2 without showing electromigration[39-41] (i.e., mass transport of the metal at high current densities) effects.

Gold and silver, because of their high atomic mobility, have a tendency to agglomerate, especially at high temperatures, making the deposition of very-low-resistance thin films difficult. Copper, although highly conducting, tarnishes rapidly in air unless protected within a few hours of deposition. The use of these materials in circuits requiring processing at temperatures in excess of 200°C is therefore not advisable. Aluminum is by far the most widely used material and represents the best compromise for interconnections, even though it is prone to scratching and corrosion. Aluminum interconnections can withstand temperatures up to 350°C. However, since it does not form very robust mechanical joints with soldered external connections, it is generally desirable to restrict the use of Al for interconnections only and use Au or Cu for contacts where soldered joints for external connections are needed. Since the adhesion of all these materials to glass is not very good, an underlayer of Cr or nichrome (Ni–Cr alloy) is used for improving their adhesion. The combination nichrome–Cu–Pd is very suitable for contacts.

In the case of semiconductor integrated circuits, in addition to the above-mentioned requirements, the material used for contacts must also form an ohmic (i.e., a low-resistance contact with linear, nonrectifying I–V characteristics) contact both with p- and n- type Si. Here too, Al has proved very successful. However, because of its low scratch and corrosion resistance, alternative systems such as Mo–Au[42] and Ti–Pt–Au[43] have been proposed. A thin (~ 0.1 μm) layer of sputtered Mo adheres strongly to Si and forms an ohmic contact with it. A 1-μm-thick layer of Au on it provides the required conducting path. The Mo–Au contact is stable up to a temperature of 370°C, above which it fails due to diffusion of Au through Mo into Si to form a Au–Si eutectic compound. Frequently, a thin layer of Pt or Al is deposited on Si prior to Mo to further ensure the ohmic nature of the contact. It has been claimed that the Mo–Au system has higher scratch resistance than Al and is less prone to electromigration effects. This system, however, does not offer additional corrosion resistance. Another alternative to Al, developed at Bell Telephone Laboratory, is the Ti–Pt–Au system. Initial contact to Si is made by depositing a thin

(~ 0.07 μm) layer of Pt to form an ohmic contact. This is followed by a thin layer of Ti (0.05 μm, by diode sputtering) which provides good adhesion to the contact, and a thin layer of Pt (0.15 μm, by diode sputtering) which serves as a diffusion barrier between the Au and Ti. A vacuum-evaporated or electroplated layer of Au serves as the conductor. The main advantage of this system is high corrosion resistance and stability up to temperatures around 350°C. Both these alternatives are, however, much costlier than the Al contact. The same materials can also be used in thin film integrated circuits.

4.3. Thin Film Active Components

Of the various thin film active elements proposed, the thin film transistor (TFT) has proved the most successful. Apart from having desirable electrical characteristics, a TFT has the advantage of being capable of fabrication at the same time and by the same techniques as thin film passive elements, thus making it possible to fabricate all-thin-film integrated circuits in a single pumpdown cycle of the vacuum system.

4.3.1. Thin Film Transistor (TFT)

A TFT[44,45] is an insulated-gate field effect transistor (IGFET) and is analogous in operation and electrical characteristics to the conventional metal–oxide–semiconductor field effect transistor (MOSFET). An FET is a type of unipolar device in which the conductance of a semiconductor region lying between two electrodes, called the source and the drain and forming low-resistance ohmic contacts with it, is modulated by means of an electric field applied to it by applying a voltage to a third electrode, called the gate, which is separated from the semiconductor by an insulator layer or by a reverse-biased p–n junction. The former type of FET is known as the insulated-gate field effect transistor (IGFET) and the latter as the junction field effect transistor (JFET). An advantage of the former over the latter is that the gate may be biased either positively or negatively with respect to the source, without drawing appreciable gate current. Figure 4.5 shows the various geometrical configurations in which such a device may be realized in thin film form. In the coplanar electrode structure (Fig. 4.5c) where the semiconductor layer is deposited before the electrodes, maximum freedom exists, during fabrication, for processing of the semiconductor to obtain desirable properties, without having the electrodes go through the same. Configurations (a) and (d) allow maximum freedom in the choice of fine pattern fabrication techniques without any possibility of damage to the semiconductor.

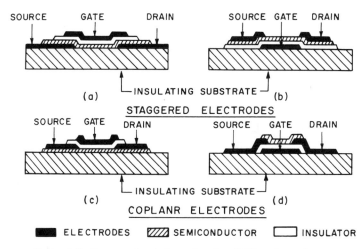

Figure 4.5. Cross-sectional views of various TFT configurations.

The first successful TFT was fabricated[46] by Weimer using CdS as semiconductor. Apart from CdS, other semiconductor materials which have proved most successful for TFTs are CdSe[47] and Te[48], although materials such as InSb,[49] SnO_2,[50] PbS,[51] PbTe,[52] and InAs[53] are also quite suitable. The insulator film is generally of SiO, although SiO_2 and CaF_2 have also been used. For making electrodes, Au is generally used when the electrodes are underlying [as in configurations (a) and (d)], and Al when they are overlying [as in configurations (b) and (c)]. To give an idea of the thickness of the various layers, the thickness of the semiconductor layer lies in the range of a few hundred angstroms to about 1 μm, that of the insulator layer lies in the range 200–2000 Å, and the electrodes are several hundred angstroms thick.

Similar to the conventional FET, a TFT is a high ($\sim 10^9 \, \Omega$) input impedance device with extremely well-saturated pentode-like characteristics. The frequency response of a TFT is almost flat from dc up to several MHz, after which it starts falling at a rate expected from the output impedance. As a switch, TFT has a very high ($\sim 10^5–10^7$) on/off conductance ratio. Figure 4.6 shows typical drain current versus drain voltage ($I_d–V_d$) characteristics for a TFT having an n-type conduction channel, in particular a CdS TFT operating with source grounded. A TFT with characteristics as shown in Fig. 4.6a is called an enhancement-type TFT because the drain current is zero at zero gate bias and a positive gate bias V_0 is required for the onset of conduction. The drain current increases further by several orders of magnitude as the gate bias is increased to several volts positive with respect to the source. Let us see why this enhancement in the conductance of an n-type semiconductor takes place

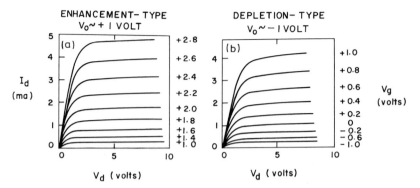

Figure 4.6. Typical drain current vs. drain voltage (I_d–V_d) characteristics of (a) enhancement-type and (b) depletion-type TFT (after Ref. 45).

by a positive gate bias. The reason is that the gate electrode forms one plate of a capacitor with the semiconductor as the second plate, so that a positive charge on the gate induces an equal amount of negative charge on the surface of the semiconductor near the semiconductor–insulator interface. Since the semiconductor is of n-type, the induced negative charge enhances its conductance in a thin surface region ($\sim$ a few tens of angstroms thick) called the conduction channel. On the other hand, in the depletion-type TFT (Fig. 4.6b), a finite useful drain current flows initially even at zero gate bias, and a negative gate bias V_0, called the pinch-off voltage, is required for reducing the drain current to zero. Obviously, a depletion-type TFT is also capable of being operated in the enhancement mode by making the gate positive. However, for IC applications an enhancement-type TFT is always preferred over the depletion type. The design parameters determining the value of V_0 will be discussed later.

It has been shown[54] by Borkan and Weimer that the drain current I_d for values of drain voltage V_d up to the knee of the characteristic curves is approximately given by

$$I_d = (\mu_d C_g / l^2)[(V_g - V_0)V_d - V_d^2/2] \qquad (4.8)$$

where μ_d is the field effect mobility of the carriers, l is the source–drain spacing, C_g is the total capacitance across the gate insulator, V_g is the gate voltage with respect to the source, and V_0 is the threshold voltage as defined earlier. The quantity V_0, which determines whether the TFT is of enhancement or depletion type, depends on some sort of a "built-in" gate bias and is given by

$$V_0 = -N_0 e / C_g \qquad (4.9)$$

where e is the electronic charge and N_0 represents the number of free carriers present at zero gate bias for a depletion-type TFT, being proportional to the semiconductor thickness and the average volume density of the carriers. For an enhancement-type TFT, N_0 is negative and represents the total number of acceptors or traps which must be filled before the onset of conduction. The derivation of expression (4.9) is based on the assumption of a V_g-invariant mobility and a semiconductor layer which is thin compared to the normal space-charge layers.

For drain voltages greater than that corresponding to the knee of the $I_d - V_d$ curve, i.e., for $V_d > V_g-V_0$, the gate becomes effectively negative with respect to the drain and produces a region depleted of carriers in the vicinity of the drain. As the drain voltage is further increased, the depletion or the pinch-off region grows in length by moving toward the source till it extends throughout the conduction channel, so that the drain current increases very little with increase in V_d beyond V_g-V_0 and almost saturates at a value given by

$$I_{ds} = (\mu_d C_g/l^2)(V_g-V_0)^2 \tag{4.10}$$

The saturation behavior of the I_d-V_d characteristics gives to the device a high output impedance and a high voltage gain. A figure of merit which is used to assess the high-frequency performance of a TFT is its gain–bandwidth product GB, defined as

$$GB = g_m/2\pi C_g \tag{4.11}$$

where g_m is the transconductance defined as $(\partial I_d/\partial V_g)_{V_d}$ and is given by

$$g_m = \begin{cases} \dfrac{\mu_d V_d}{l^2} C_g, \\[4pt] \text{in the region below the knee of the curves (that is, } V_d < V_g - V_0) \\[8pt] \dfrac{\mu_d(V_g - V_0)}{l^2} C_g, \\[4pt] \text{in the saturation region (that is, } V_d \geq V_g - V_0) \\ \text{where the TFTs are generally operated} \end{cases} \tag{4.12}$$

TFTs generally have lower values of the GB product than Si MOSFETs because of low mobility values in evaporated films. GB products of up to 30 MHz in CdS and CdSe, and up to 170 MHz in Te TFTs have been reported.[55] Typical values of g_m for CdS TFTs range from 4000 to 25,000 μmhos.

The various parameters which are of great importance for a TFT are its I_{ds}, g_m, GB, and on/off conductance ratio, and they depend on the geometrical design of the TFT through Eqs. (4.10)–(4.12); for example, I_{ds},

$g_m \propto W/l^2 t$, and GB $\propto 1/l^2$, where l is the length of the source–drain gap, W is its width, and t is the thickness of the insulator layer. We thus see that the dimensions of the gap are very critical in determining the performance of a TFT. It is therefore important that during fabrication of the source and the drain electrodes, the surface mobility of the evaporant metal atoms be kept as low as possible. Scattering of metal atoms into the shadowed regions is less with Al than with Au, and the small amounts which do get scattered in Al are very likely to oxidize. Further, the undesirable surface states due to these atoms scattered over the semiconductor are less effective if the source and drain electrodes lie on the opposite side of the semiconductor film from the gate insulator, as in the staggered electrode structures (Fig. 4.5a and b).

The maximum gate voltage which can be applied to a TFT is another important parameter and depends on the dielectric strength of the insulator material and its thickness. An optimum insulator thickness is required to make a compromise between the transconductance and the maximum permissible gate voltage. The choice of a depletion- or enhancement-type characteristic can be made by controlling the density and the type of surface states at the semiconductor insulator interface. The magnitude of V_0 required for most applications can be made small by using a thinner semiconductor and a thicker insulator layer. The presence of slow trap states at the semiconductor–insulator interface can lead to a slow drift of V_0 to higher values as in the case of MOSFETs. CdSe TFTs using Al_2O_3 insulator layers have been found[56] to be more stable than those using SiO.

Another type of thin film transistor with performance superior to the CdS TFTs is the thin film MOSFET, which employs[57] an epitaxial p-Si thin film instead of single-crystal Si, grown on an oriented, highly polished, single-crystal sapphire by pyrolytic decomposition of silane (SiH_x). The requirement of a single-crystal substrate, however, still remains a limitation.

Thin film transistors have established their usefulness as fast (switching time $\sim$ a few nanoseconds) switching elements with high ($\sim 10^5$ to 10^7) on/off conductance ratios and can therefore be used for addressing thin film display panels, for performing logic operations, in shift registers, etc. Amplifying devices with voltage amplification factors greater than 100 and input impedance greater than 1 MΩ and shunted by 50 pF can also be made. Some of the important all-thin-film integrated circuits employing TFTs are discussed in Section 4.4.

4.3.2. Thin Film Diodes

A TFT can be used as a diode simply by connecting the gate to the drain. The resulting diode characteristics are determined by the threshold voltage V_0 of the TFT, and it is preferable to have an enhancement-type TFT with V_0 close to zero. A depletion-type TFT gives rise to a large reverse saturation current, whereas a strongly enhancement type requires application of a large forward bias for onset of conduction.

Another type of diode is the Schottky-barrier diode, formed by making a blocking type of contact between a semiconductor and a metal. Such a contact is made by an n-type semiconductor with a higher-work-function metal and a p-type semiconductor with a lower-work-function metal, as can be seen from the energy-band diagrams of these systems (Fig. 4.7). It should be noted, however, that the depletion region formed at the junction should not be too thin, otherwise tunneling effects can give rise to linear I–V characteristics. A Schottky diode is generally a layered structure with an ohmic contact at the bottom and a blocking contact at the top. A frequently used diode is the CdS–Te diode with a lower ohmic contact of Au–In, rectification ratios greater than 10^4 (at 1.5 V), and reverse-breakdown voltage ~5 to 10 V. The shunt capacitance of this diode is higher than a comparable Si diode because of the thinness of the depletion region. Its performance can be improved[58] by inserting a very thin insulating layer as

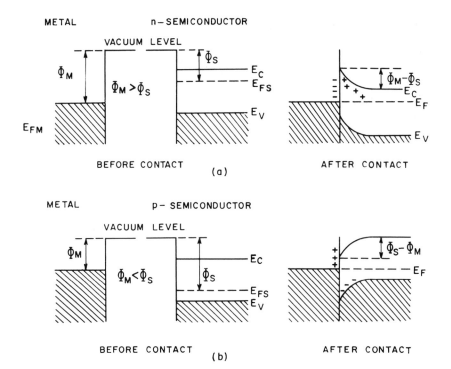

Figure 4.7. Energy band diagrams of Schottky-barrier diodes.

that of CdTe, SiO_x, or ZnS between CdS and the blocking contact. A Ti–TiO_2–Pd Schottky diode, with TiO_2 formed by anodization of a Ti film, gives[59] rectification ratios up to 10^6 at 1.5 V and reverse-breakdown voltages $\sim$ 10 to 15 V. In spite of the poorer performance of thin film diodes compared to Si diodes, their use has been successfully demonstrated in a number of all-thin-film integrated circuits, as described in the next section.

4.4. Thin Film Integrated Circuits

The possibility of having all-thin-film integrated circuits emerged with the advent of the thin film transistor, prior to which the TFICs were essentially hybrids in the sense that only the passive components were fabricated on the substrate using thin film techniques, while the active were added later on either as discrete components (by soldering or microwelding techniques) or as Si ICs. The general advantages of the TFICs in common with Si ICs are:

1. In integrated circuits, since the internal connections are already made, assembling of complicated electronic systems from ICs rather than from discrete individual components requires fewer interconnections.
2. The reduction in the number of connections gives higher reliability to systems incorporating ICs. This is so because failures in electronic systems arise largely from defects in the interconnections rather than in the components themselves, as, for example, poor joints, accidental shearing and bending of leads, and corrosion and removal of leads under the influence of heat. This higher reliability thus reduces the cost of ownership.
3. The high degree of automation used in the manufacture of ICs minimizes failures due to human error.
4. ICs are rugged structures with high mechanical and thermal shock resistance because of their small size and low physical and thermal mass.

TFIC technology is superior to Si IC technology in the following respects:

1. The parasitic losses which occur in Si ICs due to spurious coupling between the various components are absent in TFICs because of the excellent isolation provided by the insulating substrate (glass/ceramic).
2. Fabrication of TFICs generally involves only a single process (vacuum deposition) with very simple post-deposition techniques,

such as annealing and baking, carried out at temperatures much lower (~ 500°C) than those required (~ 1000°C) in Si ICs. In contrast, fabrication of Si ICs involves repeated cycles of oxidation, application of photoresist coating, photomasking, developing, etching, cleaning and diffusion, as well as vacuum deposition for contacts and interconnections. Furthermore, since each of these processes takes place in separate pieces of equipment, careful handling and cleaning is required at each step. In TFICs, on the other hand, the substrate need be handled only twice, once while inserting it in the vacuum system and another while removing it from the system.

3. TFICs are fabricated on ordinary, easily available and inexpensive glass–ceramic substrates. For Si ICs, nothing but high-purity, defect-free single-crystal Si substrate is required.

4. In TFICs, control of the various circuit parameters can be readily achieved by monitoring the thickness and electrical characteristics of the various components during deposition. Further, finer adjustments in the values of the components can be made after deposition by one of the trimming methods. No such control and adjustment is possible with the diffused components in Si ICs.

5. High-value resistors and capacitors cannot be fabricated for Si ICs using diffusion techniques. In such cases, use is made of the thin film resistors and capacitors.

Electronic circuits employing thin film components are widely used[60–64] in a number of areas. For example, thin film resistor networks are used in electronic switching systems, medical electronic equipment, ladder networks for digital-to-analog conversion, and attenuators for adjusting the transmission level of telephone toll lines. Thin film components are also used in a number of digital, analog, and microwave circuits. Tantalum nitride resistors and tantalum oxide capacitors are used to fabricate RC networks for use in tone generators for the Trimline® and Touch-tone® telephone dialing system.

The first advanced all-thin-film integrated circuit, fabricated[65] by Weimer *et al.*, was a 180-stage scan generator (Fig. 4.8). This scan generator is a modified version of a shift register and consists of 360 CdSe TFTs, 180 CdSe field effect diodes, 360 resistors, and 180 capacitors. Output voltage pulses of 2 to 4 V are obtained with clock frequencies from 12 KHz to 2 MHz. Two such 180-stage shift registers were later used[66] by Weimer *et al.*, in conjunction with a column of 180 video-coupling transistors, to scan a photosensitive array comprised of 32,400 (i.e., 180×180) picture elements with center-to-center spacing of 2.08 μm between them. Each picture element consists of a CdS (or CdS–CdSe alloy) photoconductor in series with a diode switch and is connected to two mutually perpendicular

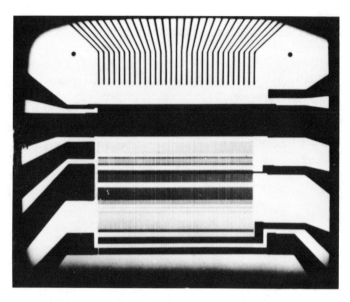

Figure 4.8. A photomicrograph of the 180-stage thin film scan generator (after Ref. 65).

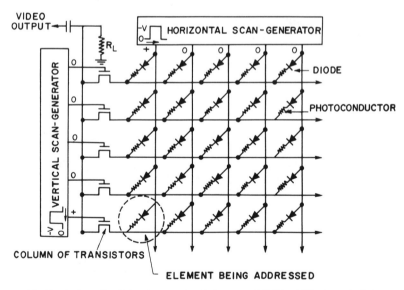

Figure 4.9. Equivalent circuit for the completely integrated (180 × 180) element photosensitive array showing the method of connecting the scan generators and coupling out the video signal (after Ref. 66).

address strips, as shown in Fig. 4.9. The shift registers were driven by square-wave pulses provided by external driving circuits. The picture elements in the array are sequentially addressed by the coincidence of voltage pulses applied by the scan generators to the horizontal and vertical (X–Y) address strips. The image is formed by charge storage as in the case of a vidicon (see Chapter 3). The array was scanned at conventional T.V. scan rates so as to allow the picture to be displayed in the receiver of a commercial T.V. Figure 4.10 shows a photomicrograph of the complete thin film camera. In addition to their use in all-solid-state T.V. cameras, these self-scanned image sensors find application in pattern-recognition devices, information-processing arrays, photoreaders, etc.

Another area where TFT is expected to play an important role is in flat-panel, matrix-addressed, liquid-crystal and electroluminescent displays. The fabrication of TFT matrix of areas as large as (14.16×14.16) cm^2 and employing 56,000 TFTs and 28,000 thin film capacitors, for use in addressing of electroluminescent displays, has been reported[67] by Fang-Chen Luo *et al.* The successful operation of TFT-driven liquid-crystal displays and electroluminescent displays has been reported by a number of workers[68–70] in this area, and the future of flat-panel T.V. appears very bright.

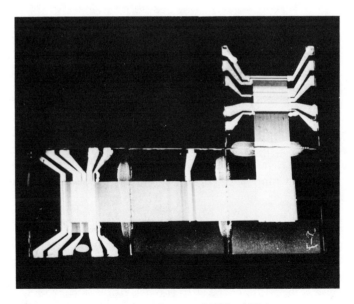

Figure 4.10. Photograph of a completely integrated (180×180) image sensor comprised of the photosensitivity array, the horizontal scan generator, the vertical scan generator, and the video-coupling circuit (after Ref. 66).

4.5. Microwave Integrated Circuits (MICs)

Microwave integrated circuits (MICs)[71,72] represent an extension of the IC technology to microwave frequencies. MICs provide a miniaturized, more reliable and cost-effective alternative to the conventional microwave circuits employing bulky individual components connected together. Since thin film components can be made much smaller than the wavelength of typical microwave signals, it is possible to have lumped-parameter passive elements in place of the distributed-parameter elements used in conventional microwave circuits where lumped-parameter elements become practically obsolete, especially at high frequencies. Similar to the conventional IC technology, MIC technology may also be either monolithic or hybrid. The substrates most suitable for monolithic circuits are semi-insulating GaAs and Si. Hybrid circuits employ dielectric substrates with passive components formed on them using thin-film–thick-film techniques and active components added to the circuit as discrete devices.

The basic structure from which the various MIC passive components can be derived is the planar transmission line, similar to the waveguide/coaxial wire transmission line of conventional microwave electronics. One simple form of the commonly used planar transmission line is the microstrip, which consists of a dielectric substrate with a narrow strip of a conductor deposited on one side and a conducting ground plane deposited on the opposite, parallel face, as shown in Fig. 4.11a. In a slightly modified version of the microstrip, called the *stripline*, the conducting strip lies symmetrically between two parallel conducting ground planes, as shown in Fig. 4.11b. The dielectric between the two ground planes may or may not be uniform.

The dominant mode of wave propagation in both the microstrip and the stripline is the TEM mode (i.e., the electric and magnetic fields lie in the cross-sectional plane transverse to the direction of propagation), as shown in Fig. 4.12 for the case of a microstrip. It is clear from the figure that most of the energy is propagated in the dielectric just below the strip. The transmission losses in the case of a stripline are low compared to a microstrip, because the fields remain confined to the region between the ground planes.

The three parameters which are of great importance to the transmission line are its characteristic impedance (Z_0), line wavelength (λ), and attenuation constant. These parameters depend[73] upon the geometry of the transmission line and the nature of the dielectric. For example, the characteristic impedance of a planar transmission line decreases with increasing w/d ratio (w is the width of the strip and d is its distance from the ground plane) and dielectric constant ε of the medium surrounding the strip. We note here that the existence of a pure TEM mode is possible only if propagation takes place in a uniform dielectric. However, since the fields

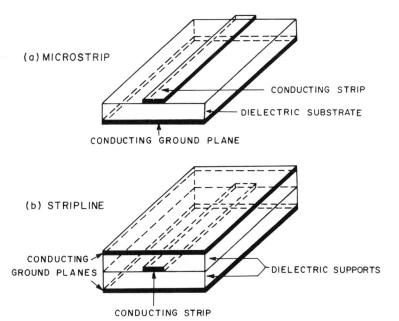

Figure 4.11. Geometry of a microstrip (a) and a stripline (b).

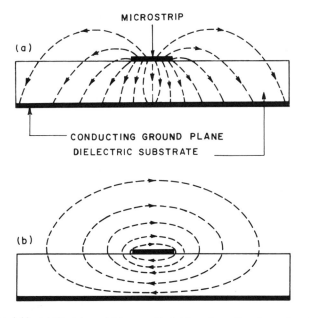

Figure 4.12. (a) Electric and (b) magnetic field configurations in a microstrip.

in a microstrip extend into air also, deviations exist from the pure TEM mode, as a consequence of which the line wavelength (λ) differs from that corresponding to the TEM mode (λ_{TEM}). The line wavelength λ decreases and approaches λ_{TEM} as (1) the strip is made wider and the dielectric thinner (i.e., as w/d increases) and (2) the dielectric constant ε is reduced. Attenuation in a planar transmission line takes place due to both conductor losses as well as dielectric losses. If the propagation takes place in a uniform dielectric, dielectric losses are independent of the geometry of the line. The conductor losses, on the other hand, decrease as the w/d ratio increases and as ε decreases. The conductor losses are also minimized if the thickness of the conductor is made at least a few times the skin penetration depth.

MIC elements, such as capacitive stubs, inductive stubs, quarter-wavelength impedance transformers, etc., for use in distributed-parameter networks, can be constructed out of the microstip by using appropriate lengths of the transmission line, in a manner analogous to the conventional microwave transmission lines. For example, a quarter-wavelength long microstrip, having characteristic impedance Z_0, acts as an impedance transformer between impedances Z_1 and Z_2, where Z_1 and Z_2 are related to Z_0 by the equation $Z_0 = (Z_1 Z_2)^{1/2}$. Very short sections of microstrip (shorter than approximately one-eighth the line wavelength) can also be used as lumped-parameter resistors, capacitors, and inductors. The various circuit components are delineated on a continuous metal film using lithographic techniques. Use of ferrimagnetic substrates permits the fabrication of a variety of nonreciprocal components like circulators, isolators, switches, and phase shifters.

We have not gone into the details of the various MIC components and circuits because their basic understanding requires a thorough understanding of conventional microwave electronics, a discussion of which would be out of place in this book. It is sufficient here to say that thin film integrated circuit techniques are currently being applied successfully to the fabrication of a number of microwave circuits. For details, we refer the interested reader to the extensive literature available[71,72] on MICs.

4.6. Surface Acoustic Wave (SAW) Devices

4.6.1. Introduction

Devices employing surface acoustic waves (SAW) for analog processing of electrical signals in the microwave range (a few megahertz to a few gigahertz) are known as SAW devices. Surface acoustic waves (or Rayleigh waves, as they are known after their discoverer, Lord Rayleigh) are elastic displacements propagating nondispersively (i.e., with velocity independent of frequency) at the stress-free boundary of a solid, with the amplitude decaying exponentially in the bulk with increasing distance from the surface. For the case of a semi-infinite, homogeneous, and elastically isotropic solid, the SAW consists of elliptical particle motion which can be resolved into two orthogonal components: one in the direction of propagation (longitudinal wave) and the other normal to the free surface (shear wave). The ellipse of motion thus lies in the sagittal plane (a plane containing the wave vector and perpendicular to the surface). The velocity of the Rayleigh wave depends directly on the elastic stiffness constants of the medium and inversely on the square root of its mass density. In the case of an elastically anisotropic medium such as a piezoelectric, the particle motion, in general when the sagittal plane is not a symmetry plane, may have an additional third orthogonal component in the plane of the surface. This implies that the elliptical particle motion is not necessarily confined to the sagittal plane, so that power flow is not parallel to the direction of wave propagation, except along selected pure mode axes. Obviously, the SAW velocity in an anisotropic medium becomes anisotropic. If, in addition, the anisotropic material is also piezoelectric, the SAW is accompanied by a traveling electric field (electroacoustic waves) both inside and outside the solid, thus making it possible to launch as well as to detect acoustic waves simply by depositing suitably structured thin film metal electrodes on the surface. The SAW velocity, in the case of an anisotropic material, depends also on the piezoelectric and dielectric properties of the medium and is higher than that for an elastically similar but nonpiezoelectric material (piezoelectric stiffening).

Since the velocity of surface acoustic waves is approximately five orders of magnitude smaller than that of electromagnetic (em) waves, SAW devices are 10^5 times smaller than the corresponding conventional devices using em waves. Further, the attenuation of SAW (typically ~ 1.5 dB for 10^4 wavelengths) is orders of magnitude smaller than that for em waves of the same frequency. The propagating SAW is always accessible and can be readily manipulated by depositing suitable thin films on the surface. This has given birth to a versatile microminiaturized technology for analog processing of em signals. Some of the notable devices include simple and multiple-tapped delay lines, band-pass filters, pulse-compression filters, frequency and phase coders and detectors, frequency mixers, spectrum analyzers, Fourier transformers, convolvers, waveguides, amplifiers, resonators, oscillators, etc. The most attractive areas of application of these devices are in radar and sonar communication systems and in color T.V. and displays.

SAW devices are rugged, planar devices, fabricated using thin film technology either on single-crystal piezoelectrics or on polycrystalline oriented/epitaxially grown single-crystal piezoelectric thin films of thickness less than one SAW wavelength, deposited on nonpiezoelectric substrates. The requirements on the material for SAW devices are high electromechanical coupling (i.e., the value of the electric field at the surface per unit power in the SAW), low attenuation and high thermal stability. The materials most commonly used for commercial SAW devices are LiNbO$_3$ and ST-cut quartz (as single crystals) and ZnO (in the form of thin films deposited by sputtering). Whether the SAW devices are made on single crystals or on thin films, the basic principles involved are the same. In the present section, we will discuss how various signal-processing functions can be accomplished in these devices using thin film overlays.

4.6.2. SAW Transducer

A SAW transducer is a device which converts electrical signals into surface acoustic waves and vice versa. The simplest type of a transducer consists of a piezoelectric film or substrate with an alternate phase array (also called the interdigital comb) of electrodes, as shown in Fig. 4.13a. To generate a SAW in an interdigital transducer (IDT), an alternating voltage is applied between the adjacent sets of fingers, which are separated by a distance d equal to half the SAW wavelength (λ) at the design

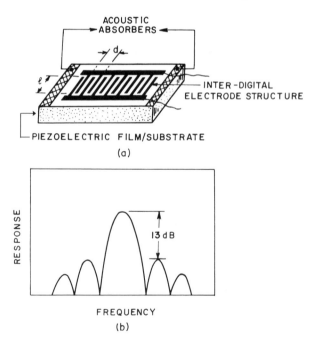

Figure 4.13. (a) A SAW transducer with interdigital electrodes; (b) frequency response of the above transducer.

frequency. Increasing the number of finger pairs increases the efficiency of the transducer to a certain extent, at the cost of an accompanying decrease in the bandwidth, so that an optimum number of electrode finger pairs is required to obtain a maximum efficiency–bandwidth product. This is determined by the phase velocity of the SAW, which depends on the properties of the piezoelectric material and its orientation and thickness (in the case of thin films). The electrode dimensions, that is, the active length l of the fingers (the amount adjacent fingers overlap) and the periodic spacing d between them, determine the frequency response of the transducer. For the simplest version, when the finger lengths are equal and the finger spacing is uniform, the frequency response is a $\sin x/x$ curve (Fig. 4.13b), with the primary side lobes 13 dB below the main peak.

The transducer described above is bidirectional in the sense that it produces waves both in the forward as well as in the backward direction (one of them can be terminated by using an acoustic absorber, such as wax or tape, at the end), leading to a decrease of 3 dB in the conversion efficiency of the transducer. Unidirectional transduction can be achieved by using two identical IDTs separated by a distance $(n + \frac{1}{4})\lambda$ (where n is an integer). The IDTs are driven 90° out of phase using either two separate generators or a single generator with a $\lambda/4$ transmission line connecting the two IDTs.

The maximum frequency of the SAW in the fundamental mode that can be generated using IDT is limited by the minimum periodic spacing (which for a 100-MHz transducer on $LiNbO_3$ is typically $\sim 9\ \mu m$) that can be obtained using lithographic techniques. Transducers with operating frequency up to 3.5 GHz in the fundamental mode have been made[74] using electron-beam lithography. The frequency response of an IDT can be modified by adjusting the finger spacings in order to control the frequency characteristics, and the active finger lengths (apodization) in order to control the amplitude weighting of the characteristics. For example, continuous grading of the pitch of the electrode array results in broad-band characteristics. Variation of the finger lengths of one of the electrodes provides heavier weighting of the midband frequencies. Thus, it is possible to design and fabricate an IDT with a desired frequency response.

IDTs are used in delay lines and filters for generating, detecting, and tapping surface acoustic waves. The desired signal-processing function can be accomplished by a proper choice of the location and frequency response of the IDT, as discussed in the following sections.

4.6.3. SAW Delay Line

Two IDTs, one input and one output, on a single piezoelectric surface separated by some distance form a delay line (a device for introducing a predetermined value of time delay in the signal). A broad-band delay line

(a) DISPERSIVE DELAY LINE

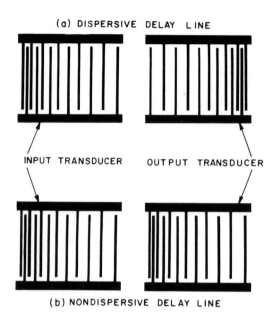

INPUT TRANSDUCER   OUTPUT TRANSDUCER

(b) NONDISPERSIVE DELAY LINE

Figure 4.14. Electrode geometry for a dispersive (a) and a nondispersive (b) delay line.

using two IDTs with varying periodicity can be made dispersive or nondispersive by placing the transducers in the configurations shown in Fig. 4.14a and b, respectively. In configuration (b), the different-frequency acoustic waves travel equal distances on the delay line and therefore suffer equal delays, making the line nondispersive. This is not so in configuration (a), and thus the delay line is dispersive. Use of bidirectional transducers in delay lines imposes an insertion loss of 6 dB. A delay line employing unidirectional IDTs exhibits reduced losses and good triple-transit suppression (the ratio, in decibels, of the amplitude of the direct delayed signal to the amplitude of the first multiply reflected echo received at the output transducer). Delay lines find application in radar and sonar communication and in computers.

4.6.4. SAW Band-Pass Filter

A common configuration of a band-pass filter is shown in Fig. 4.15a. It consists of an input single-finger-pair wideband transducer and an output transducer in which the finger spacing is uniform but the active finger length varies with distance as $\sin x/x$. The voltage generated at the output IDT as the acoustic pulse moves across it (or the impulse response of the IDT) is given by the spatial image of the IDT structure and therefore has a $\sin t/t$ dependence, as shown in Fig. 4.15b. The frequency response of the filter, which is just the Fourier transform of the impulse response, is therefore

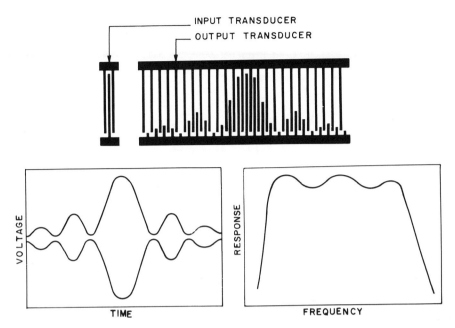

Figure 4.15. A SAW band-pass filter (a) with corresponding impulse response (b) and the frequency response (c).

almost rectangular (Fig. 4.15c), giving it band-pass characteristics. Such band-pass filters are commercially used[75] in video IF stage in color T.V. Typically,[76] an Al-deposited-on-quartz band-pass filter designed at a central frequency of 25 MHz has a bandwidth of ~ 2 MHz and a rejection of ~ 21 dB.

4.6.5. *SAW Pulse-Compression Filter*

A pulse-compression filter is a device which outputs an energy pulse of narrow time duration with minimum spurious response, when an input coded signal (generally chirped, i.e., linearly frequency modulated) having a substantially longer time duration is fed into it. Such filters find application in high-resolution radar systems.

A pulse-compression filter consists of a single-finger-pair wide-band transducer at the input and a multiple-finger-pair transducer with constant active-finger length but monotonically decreasing finger spacing at the output, as shown in Fig. 4.16a. The left-hand side of the output IDT having wider finger spacings responds to lower frequencies, and the right-hand side having narrower finger spacings responds to higher frequencies. Let a constant-amplitude, chirped RF voltage pulse, in which the frequency

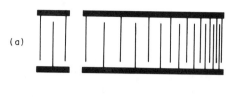

(a)

(b)

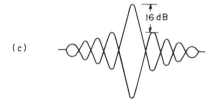

(c)

Figure 4.16. Schematic representation of a SAW pulse-compression filter (a) for chirped signals (b). The impulse-response (or the actual output voltage generated as the acoustic pulse travels along the filter) is shown in (c).

decreases linearly with time, be applied to the input IDT. As the SAW generated by it reaches the output IDT, the low-frequency region at the entrance does not respond well to the leading high-frequency part of the SAW pulse. However, as the SAW pulse travels further, the leading high-frequency part of the pulse falls at the high-frequency-responsive region of the IDT and the trailing low-frequency part falls at the low-frequency-responsive region of the IDT, so that the response of the filter is very high. After some time, as the pulse moves further to the right, the trailing low-frequency part of the pulse lies on the high-frequency-responsive region of the IDT and, therefore, once again the response of the filter is low. The actual impulse response (output voltage versus time) has the input-pulse center frequency with a $\sin \alpha t/t$ modulation, as shown in Fig. 4.16c. The input chirped pulse is thus compressed into the narrower central lobe of the $\sin \alpha t/t$ output, with the primary side lobes 16 dB below the main peak.[77]

Use of an apodized output IDT, in which the active finger length and hence the coupling at each frequency tap varies (in addition to the monotonically decreasing finger spacing) as $\cos^2 x$ with distance, reduces the side lobes in the output of the pulse-compression filter, at the expense of a tolerable increase in the width of the central lobe.

4.6.6. SAW Amplifier

Losses occurring in acoustic delay lines due to intrinsic attenuation (due to phonon–phonon collisions), reflections, scattering, etc. of the SAW become prohibitive at very high frequencies or when the delay line is designed for introducing very long delays. Amplification of the SAW is therefore required.

A SAW amplifier is based on the principle that when the electric field associated with a SAW propagating in a piezoelectric material interacts with the charge carriers drifting along with it under the action of an applied electric field, with a velocity greater than that of the SAW, energy is delivered from the charge carriers to the propagating SAW. As a consequence of this, the amplitude of the SAW increases with distance as it travels, and amplification results.

There are two types of SAW amplifiers: monolithic amplifiers and film amplifiers. In the former type, the piezoelectric material is also a semiconductor so that the interaction of SAW takes place with the charge carriers present in the material itself. Monolithic amplifiers can be fabricated using materials such as CdS, CdSe, and ZnO. A film amplifier consists of a thin film of a semiconductor (or a piezoelectric) on a piezoelectric (or a semiconductor) substrate, for example, InSb/LiNbO$_3$, ZnO/Si, and LiNbO$_3$/Si. A schematic diagram of a film amplifier is shown in Fig. 4.17.

4.6.7. SAW Guiding Components

A wide variety of two-dimensional signal-processing devices is made possible by control of the surface acoustic waves by guiding, focusing, and reflecting. A path of low phase velocity of SAW acts as a guide for the propagating waves. Waveguides on nonpiezoelectric materials are formed by: (1) depositing a thin film of a denser and less stiff material than the substrate (loading layer, e.g., Au/fused silica) on the path where guiding is required, or (2) depositing a thin film of stiffer and less dense material

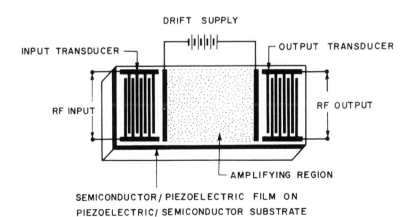

Figure 4.17. SAW film amplifier.

than the substrate (stiffening layer, e.g., Al/T-40 glass) in regions other than the guiding path. The former type is called the *slow-on-fast* waveguide and the latter the *fast-on-slow* waveguide. In the case of a piezoelectric material, however, any conducing strip deposited on it in the guiding region can guide surface acoustic waves. Guiding takes place in these regions because the conducting film shorts out the piezoelectric fields, resulting in lower SAW velocity in the coated regions than anywhere else on the surface. It may also be mentioned here that SAW propagation in layered structures is dispersive. This is so because at low frequencies the particle motion extends far into the underlying substrate, as a result of which the SAW velocity is higher, approaching that on the bare substrate. At high frequencies, on the other hand, the particle motions remain confined to the overlay and therefore the SAW velocity is lower, approaching that on the surface of the layer material.

Propagating SAW can be focused or defocused by structures analogous to an optical lens. A SAW lens is simply a lens-shaped region on the surface, where the SAW velocity is different from that on the surrounding regions. Such a lens may be formed by (1) depositing a lens-shaped stiffening layer (for a nonpiezoelectric material), (2) depositing a lens-shaped metallic film (for a piezoelectric material), or (3) forming a lens-shaped depression or protuberance on the surface, so as to alter the SAW transit time instead of the phase velocity. Reflection of surface acoustic waves can be brought about by introducing discontinuities in the SAW velocity or in the surface contour.

4.6.8. Other Applications

In addition to signal-processing devices for communications, surface acoustic waves may also be utilized to make scanning systems and light and electron-beam modulators and deflectors for use in displays and self-scanned image sensors. Kaufman and Foltz[78] have described a self-scanned optical scanner using surface acoustic waves. It converts a spatially modulated optical pattern into an amplitude-modulated electrical output. A photoconductor film is used in conjunction with the piezoelectric to control the electric field applied to each piezoelectric element, in accordance with the intensity of illumination there. The surface acoustic waves generated represent the optical signal.

An electron-beam modulator reported by Blackledge and Kaufman[79] employs the electric field generated by a SAW propagating in a piezoelectric material to modulate electron emission from a photocathode.

In this section, we have briefly acquainted the reader with the role played by thin films in the fabrication of SAW devices. For a fuller account of these devices, the reader is referred to important review articles[80-82] and the references cited therein.

4.7. Charge-Coupled Devices (CCDs)

4.7.1. Introduction

Charge-coupled devices, or CCDs[83] as they are known, are a class of charge-transfer devices (CTDs) for use in information storage and processing, delay, time-axis conversion, and scanning purposes for displays and image sensors. The concept of charge transfer was first introduced by Sangster and Teer[84] in 1969 in the form of bucket-brigade devices (BBDs). In a BBD, a quantity of electrical charge is moved step by step through an array (or mosaic) of capacitor storage elements connected in parallel through transistor switches. BBDs were soon followed by another type of CTDs, called charge-coupled devices (CCDs), first announced by Boyle and Smith[85] in 1970. At present CCDs have completely overtaken the BBDs because of their higher charge-transfer efficiency, lower dark current and noise, and capability of higher packing densities. However, the processing technology required to achieve the full capabilities of CCDs is far more complicated than for the BBDs. In this section, we briefly discuss the principle of CCDs and their applications.

4.7.2. Principle

This new concept of charge transfer consists of storing minority carriers (representing the signal) of a semiconductor in spatially defined potential wells formed at the surface of a homogeneous semiconductor and transferring them from one well to the other by moving the potential minima. A potential well can be created at a Si surface by forming an MOS capacitor, and the potential minimum can be moved by applying the appropriate voltage to the metal electrode (or the gate) of the capacitor. In the case of an MOS capacitor formed on p-Si, application of a positive step voltage to the gate results in repulsion of majority carrier from the Si–SiO$_2$ interface and hence formation of a depletion region of negatively charged acceptor states near the surface of Si. A potential well for minority carriers can thus be created by applying a large positive step voltage to the gate, and its depth can be varied by varying the magnitude of the gate voltage. The potential well exists for time durations shorter than the time τ taken by the well to fill the charge thermally. Obviously, the voltages should be applied and manipulated in time less than τ ($\sim$ a few seconds).

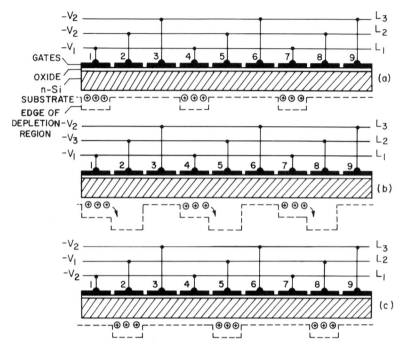

Figure 4.18. Schematic diagram showing charge-transfer action in a three-phase scheme on n-Si.

A CCD consists of a series of closely spaced MOS capacitors, in which the stored charge is transferred from one well to the other by making the voltage at the adjacent gate more attractive than that where the charge lies originally. Since the gate voltages are such that they repel the majority carriers, recombination is avoided and the charge can be shifted over long distances without undue loss.

In actual practice, charge transfer can be realized either in a two-phase or a three-phase (every second or third gate electrode connected to the same clock voltage through a common conductor) scheme. Charge transfer action in a three-phase scheme on an n-type Si substrate is shown in Fig. 4.18. The surface potential variation along the Si–SiO$_2$ interface when one applies initially to the L$_1$-clock line a voltage $-V_1$ more negative than the voltage $-V_2$ to the other two (V_2 being sufficiently negative to cause a depletion region) is shown in Fig. 4.18a. The dashed line represents the edge of the depletion region. Illumination of the device with photons to which the CCD material is sensitive generates electron–hole pairs and causes the minority carriers to be accumulated in the potential wells under the gates connected to the L$_1$-clock line. Charge may also be generated electrically by a forward-biased p–n junction formed at the input end of the device. Application of a voltage $-V_3$ ($V_3 > V_1$) to the L_2-clock line causes the potential wells under the gates connected to it to go deeper than those under the gates connected to the L_1-clock line (Fig. 4.18b). As a consequence of this, the charge stored in the L_1-potential wells spills into the L_2-potential wells. The voltages of the L_1 and L_2 lines are then changed to $-V_2$ and $-V_1$, respectively. The charge, which has now moved one spatial position, resides in the potential wells under the gates connected to the L_2 line. A similar process moves the charge further. Finally, the transferred charge may be detected at the output end by a reverse-biased p–n junction or a Schottky barrier. The MOS capacitors must lie close enough so that their depletion regions overlap and the surface potential in the regions between them varies smoothly from one capacitor to the other.

4.7.3. Applications

A large number of functions, such as shifting of information, scanning, logic, delay, etc., can be accomplished using CCDs. A CCD shift register is nothing but the structure shown in Fig. 4.18 with the addition of a charge generator at the input end and a charge detector at the output end. One of the most important contributions of CCDs is in the field of imaging.

Prior to the advent of CCDs, self-scanned image sensors utilized some sort of X–Y addressing scheme in which each picture element (consisting of a photosensitive element in series with a switch) of the array is located at the intersection of two mutually perpendicular conductor lines and is addressed by a switch when the pulses applied to the horizontal and vertical lines by scan generators (which may be a parallel-output shift register, a decoder switching circuit, or a tapped delay line) located at the periphery of the array coincided at that element. No doubt, a self-scanned image sensor employing the X–Y addressing scheme is far superior to the conventional electron-beam-scanned sensors because of its small size, all-solid-state ruggedness, low power consumption, longer lifetime, and freedom from problems encountered with vacuum tubes and electron-beam devices. Nevertheless, the usefulness of self-scanned image sensors employing the X–Y addressing scheme is limited because of some inherent problems associated with this scheme. The first problem is the relatively large capacitance associated with the readout lines compared to the capacitance of the photosensitive charge-storage elements. This limits the sensitivity of the device, especially in large-area image sensors. Another problem is the coupling of the clocks to the output at each element. Small variations in the pulses applied to the address lines are mixed with the signal and cause striations in the picture.

These problems associated with the X–Y addressing scheme are absent in image sensors using CCDs for scanning.[86,87] An optical image is focused on the substrate side of the CCD. A corresponding charge image is formed by the creation of electron–hole pairs and readout via shift-register action. Of great importance are CCDs sensitive in the IR region, which permit the fabrication of IR cameras. Such materials as InSb and GaAs are seriously being considered for IR CCDs.

A CCD display device utilizes a process that is the inverse of reading the information via shift-register action and then reverse biasing the MOS capacitors to force the minority carriers into the bulk and cause radiative recombination with the majority carriers present there.

4.8. Thin Film Strain Gauges

A strain gauge is a device which utilizes piezoresistive properties (i.e., change in resistance due to strain) of a material to measure pressure. The change in resistance takes place both due to change in the dimensions of the resistive element as well as change in the resistivity of the material. The latter change occurs because of the effect of strain on the electron mean free path. A conventional strain gauge uses a metal wire or foil as the resistive element. The change in resistance can be measured by inserting the gauge in one of the arms of a Wheatstone's bridge. These strain gauges have to be bonded onto the surface where measurement is required. The use of glue can limit both the accuracy of measurement, due to incomplete transmission of strain to the gauge, as well as the maximum temperature at which the device may be used. These limitations can be overcome in a thin film strain gauge,[88] which consists of a thin film of a piezoresistive material deposited directly on an insulating surface of the strainable device. Contacts to the strain-sensitive resistors are made by thick, low-resistance electrodes of a relatively strain-insensitive material such as Au and Al. The additional advantages of thin film strain gauges are their small size, ruggedness, and high reliability.

The resistance of a self-supporting film of thickness t, width w, and length l is given by

$$R = \rho l / A \tag{4.13}$$

where ρ is the resistivity of the film material and $A = wt$. The strain coefficient of resistance, also called the *gauge factor* (GF) and defined as

$$\gamma = \frac{dR/R}{dl/l} \tag{4.14}$$

is therefore given by [from Eq. (4.13)]

$$\gamma = 1 + 2\mu + \frac{d\rho/\rho}{dl/l} \qquad (4.15)$$

where $\mu = -\frac{1}{2}[(dA/A)/(dl/l)]$ is the Poisson's ratio for the film material. Taking into account the effect of substrate elasticity in the case of a thin film, we write expression (4.15) as

$$\gamma = 1 + \mu_s + \frac{\mu(1 - \mu_s)}{1 - \mu} + \frac{d\rho/\rho}{dl/l} \qquad (4.16)$$

where μ_s is the Poisson's ratio for the substrate material.

The material of the strain-sensitive resistor may be a metal, alloy, semiconductor, or a cermet. The actual choice is determined by the characteristics desired. The resistive and piezoresistive properties of thin films are very sensitive to their structure (see Section 4.2) and therefore to the deposition conditions. With the exception of a few magnetic materials, most of the metal films exhibit qualitatively similar piezoresistive behavior. The variation of resistance with strain is almost linear, with positive gauge factors which vary with film thickness as discussed in the next paragraph.

Theoretical and experimental studies[89,90] show that ultrathin, discontinuous metal films having high sheet resistance R_s can possess gauge factors one to two orders of magnitude higher than the bulk value (~ 2) in conventional wire or foil strain gauges. High gauge factors arise because of the strong dependence of the intergrain separation on strain. Although discontinuous films may seem very attractive for strain gauges because of their high GF, they are not generally used because of the associated problems of reproducibility and stability. The GF of relatively thick, continuous films (low R_s) approaches the bulk values. For films of intermediate thickness, where strain-insensitive surface-boundary scattering dominates, the GF assumes very low values, so that the GF vs. thickness variation shows a minimum in the curve at intermediate thickness values.

A high value of GF is not the only criterion for selecting a suitable material for fabrication of strain gauges. In addition, the material must also possess low TCR values and exhibit excellent thermal and temporal stability, so that any change in resistance due to temperature variation or temporal drift does not interfere with the measurements. The piezoresistive properties of different classes of materials are summarized in Table 4.3. In particular, metals like Pt and Cr, cermets like 70% Cr–30% SiO and 80% Au–20% SiO, and semiconductors like Si and Ge are of great interest. Some of the applications of strain gauges are in displacement

Table 4.3. Various Classes of Materials and Their Characteristics Relevant to Strain Gauges

Type and class of film material	Gauge factor γ	Temperature coefficient of γ (ppm/K)	Temporal stability
Continuous, metals	2	400	Fair
Discontinuous, metals	Up to 100	~ -1000	Poor
Continuous, metal alloys	2	~ 100	Good
Discontinuous, metal alloys	Up to 100	~ 1000	Poor
Cermets	Up to 100	200–1000	Generally good
Semiconductors	Up to 100	~ -1500	Good in some cases

transducers, torque indicators, anemometers, microphones, hydrophones, accelerometers, load cells, pressure gauges, and flow meters.

4.9. *Gas Sensors*

The strong sensitivity of the electrical properties of thin films to ambient gases, due to absorption effects, can be exploited for gas detection. Thin film NO_x gas sensors using RF-sputtered SnO_x ($x \sim 1.85$) films have already been reported.[91] These sensors show very high sensitivity to NO, NO_2, and their mixtures. For example, the resistance of the film changes linearly from $10^5 \, \Omega$ for 5 ppm of NO_x in air to $10^6 \, \Omega$ for 100 ppm of NO_x in air; the resistance in air is $10^4 \, \Omega$. These films[92] are also sensitive to the presence of H_2 and H_2S and can therefore be used for the detection of H_2 and H_2S. The performance of these films as H_2 sensors can be improved[93] by selective diffusion of gold into the grain boundaries.

MOS capacitors with transition-metal gates, such as those of Pd, Pt, and Ni, are another class of gas sensors.[94] These have been shown to be sensitive, stable and reproducible detectors of H_2, CH_4, C_4H_{10}, and CO gases. For example, the 1 MHz–10 Hz C–V characteristics of a Pd-gate MOS capacitor change[95] by -1240 and -215 mV when exposed to H_2 gas at 760 and 2×10^{-8} Torr, respectively. It has been established that this change is due to a change in the work function of the metal gate upon gas adsorption.

At present a considerable amount of effort is being directed toward the development of different types of electronic chemical sensors, giving rise to a new class of devices known as CSSDs (chemically sensitive semiconductor devices).

5

Magnetic Thin Film Devices

The present chapter is devoted to applications based on the properties of magnetic thin films. The material of this chapter has been organized into two main sections. In the first section, the reader is introduced to the subject of magnetism in thin films. The discussions in this section are limited to the extent they are relevant to device fabrication. For further details on magnetism in thin films, references[1-5] are given at the end of the chapter. The second section describes various devices.

5.1. Magnetic Thin Films

5.1.1. Introduction

Ferromagnetism in a material is a cooperative phenomenon, arising from the ordered alignment of net electronic spins due to an internal field caused by a quantum-mechanical exchange interaction. A ferromagnetic material is, therefore, characterized by a nonzero value of spontaneous magnetization (the magnetic moment per unit volume) for temperatures below a certain critical value, T_c, known as the *Curie temperature*. Above T_c, the ordered alignment of spins is destroyed due to random thermal motion, and the material behaves as a paramagnetic material. The value of spontaneous magnetization, in general, is much smaller than the characteristic saturation value M_s at a given temperature and can be increased to M_s by the application of a magnetic field (curve Oa in Fig. 5.1). This suggests in the material the presence of small ferromagnetic regions, called *domains*, each with magnetization M_s but oriented in different directions so that the resultant vector sum of magnetizations is much smaller than the saturation value corresponding to a single domain structure. These magnetic domains are separated by narrow transition regions, called *domain walls*. Magnetization of the sample upon application of a magnetic field involves (1) growth of the domains with magnetization direction closest to the applied field direction, at the expense of those with other magnetization

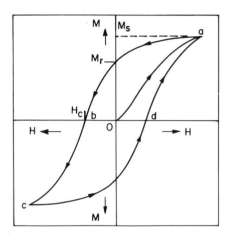

Figure 5.1. *M–H* hysteresis loop.

directions, by reversible domain wall motion at low fields and irreversible domain wall motion at high fields, and (2) rotation of spins for alignment along the field, at very high fields. Upon removing the field, the magnetization curve exhibits a hysteresis behavior (Fig. 5.1) tracing the path *ab* and retaining a finite value, M_r, of magnetization, called the *remanence* or *retentivity*. The coercivity H_c is the field required to completely demagnetize the sample.

The multidomain structure mentioned above arises as a natural consequence of the process of minimization of the total free energy of the system, which consists of (1) the exchange energy, (2) the magnetostatic or demagnetization energy, (3) the domain-wall energy, and (4) the anisotropy energy. The magnetostatic fields giving rise to the demagnetization energy arise from any discontinuity in magnetization, as for example, physical boundaries. Domain formation reduces the magnetostatic energy and increases the domain-wall energy and the exchange energy. The anisotropy energy E_k arises when the magnetization of the sample depends on its orientation with respect to a fixed set of axes. There may be several factors giving rise to the anisotropy energy, for example, (1) the anisotropic geometric shape of the sample, (2) the anisotropic magnetostriction (the phenomenon of development of strain in a material when a magnetic field is applied), and (3) the crystal structure of the material if it is in single-crystal form. The magnetocrystalline anisotropy energy averages to zero for a polycrystalline material with randomly oriented crystals. The domain and domain-wall configuration in the material are determined by the relative values of the various types of energies in the minimum energy state.

An antiferromagnetic material differs from a ferromagnetic one in that it consists of two interpenetrating sublattices of ferromagnetic spins of equal magnitude in opposite directions so as to form a stucture in which the neighboring spins tend to align antiparallel instead of parallel as in ferromagnetic materials. An antiferromagnetic material, therefore, does not exhibit a spontaneous magnetization although an ordered spin structure exists. If, however, the two opposite spins are of unequal magnitude, a finite value of spontaneous magnetization exists and the material is said to be ferrimagnetic.

5.1.2. Uniaxial Anisotropy (UA)

By virtue of their thinness, thin films are generally magnetized in the plane of the film, unless there is a large magnetocrystalline energy giving rise to an easy axis normal to the plane of the film. The magnetization vector remains confined to the plane of the film because of the very high ($\sim10^4$ Oe) demagnetization field ($=4\pi M$) perpendicular to the plane of the film. There is no preferred direction in the plane of the film. However, a uniaxial anisotropy can be induced in the film if during deposition a unidirectional magnetic field is applied in the plane of the film, resulting in a single domain configuration with easy axis parallel to the applied field. This anisotropy is called *magnetization-induced uniaxial anisotropy*. Uniaxial anisotropy of thin films is the most important property for device fabrication and is characterized by a directional dependence of energy of the form

$$E_k = K_u \sin^2 \theta \qquad (5.1)$$

where K_u is the UA constant and E_k is the anisotropy energy for a magnetization vector making an angle θ with the direction of the applied field. It can be seen from Eq. (5.1) that there are two stable states of zero energy ($\theta = 0$ and $\theta = \pi$) corresponding to the magnetization vector lying parallel and antiparallel, respectively, to the applied field. At $\theta = \pi/2$ and $\theta = 3\pi/2$, halfway between the two energy minima, two energy maxima exist which oppose any attempts to reverse the direction of magnetization from one stable state to the other. The axes corresponding to the minimum E_k and the maximum E_k are called, respectively, the *easy axis* (EA) and the *hard axis* (HA).

The origin of the above mentioned magnetization-induced anisotropy is not very clearly understood. It has been established however, that it involves at least two distinct mechanisms. According to the first, UA is induced by directional pair-ordering[6] similar to that observed during magnetic annealing of bulk specimens. The second mechanism[7] involves contributions due to the film being magnetostrictive. Strain is frozen into the film due to the process of nucleation and growth of any vapor-deposited film. The later application of a magnetic field in a different direction requires a strain to be set up in the film as M rotates from its original direction. But since the substrate is rigid, a stress is produced which leads to energy being stored in the film. Since this energy depends on the orientation of M, it is a form of anisotropy energy. Note that UA in the plane of the film may also be induced[8] by uniaxial rubbing of the substrate. Up to now we have considered only planar UA. Perpendicular anisotropy with easy axis perpendicular to the film plane can be induced under certain conditions. This is characterized by

$$E_\perp = -K_\perp \sin^2 \Phi, \qquad K_\perp < 0 \qquad (5.2)$$

$K_\perp > 0$ implies that the film normal is a hard axis while the film plane is the easy plane. Here $K_\perp$ is the perpendicular anisotropy constant and Φ is the angle between M and the film normal. When films are deposited with the vapor beam at normal incidence to the substrate, $K_\perp$ is positive with a very high value $(= 2\pi M^2)$ and therefore the EA lies in the plane of the film. However, when the films are deposited with the vapor beam at oblique incidence to the substrate, EA may take out-of-plane positions. Oblique-incidence anisotropy is induced in films in the absence of any magnetic field or in a rotating magnetic field due to structural effects. For very small angles of incidence with the film normal, planar UA is induced due to formation of "chains" of crystallites along a particular direction by a self-shadowing mechanism. But when the oblique-incidence angle is large, the crystallites tend to elongate in the direction of the vapor beam, thus forming columnar structures which give rise to high negative values of $K_\perp$ and hence perpendicular anisotropy. As expected, oblique-incidence effects are not observed in electrodeposited or sputtered films because the incoming atoms are randomized in these processes.

Perpendicular anisotropy with EA normal to the film plane is induced in normally deposited films when a large negative value of the magnetocrystalline anisotropy constant $(K_\perp)_s$ exists perpendicular to the film such that $(K_\perp)_s > 2\pi M^2$ and therefore the net value of $K_\perp = 2\pi M^2 + (K_\perp)_s$ is negative. In the case where $(K_\perp)_s$, which tends to maintain M normal to the film plane, is not sufficiently negative, so that the net value of $K_\perp > 0$, but $< 2\pi M^2$, an EA normal to the film plane is induced only when the film thickness is above a certain critical value. The importance of in-plane and perpendicular anisotropy of thin films will be illustrated in later sections of this chapter.

5.1.3. Domains and Domain Walls

The picture of a uniaxially magnetized thin film as a film where all the spins are perfectly aligned to the applied field direction is only an ideal one. In real films, deviations from this idealistic picture exist. In addition to small quasi-periodic deviations from the direction of magnetization, called *magnetization ripple* (or dispersion), magnetization in films may even split into parallel and antiparallel domains under certain conditions, in an attempt to reduce the magnetostatic energy due to free poles at the edges. Domains in thin films are separated by vertical walls perpendicular to the plane of the film. Horizontal walls parallel to the plane of the film are not possible in films of general interest because their thickness itself (~ 1000 Å) is much smaller than the wall width ($\sim 10{,}000$ Å). Application of a magnetic field in the plane of the film, for a film having planar EA, magnetizes the sample to saturation, producing a single domain structure (subject to ripple) by domain-wall motion.

The width of a domain wall and the spin configuration within it are determined by the condition of minimization of the sum of the exchange energy, the anisotropy energy, the domain-wall energy, and the magneto-static energy. An increase in wall width reduces the exchange energy but increases the anisotropy energy and the domain-wall energy.

Two types of spin configurations are possible within a wall. In one the spin changes from one direction to the other by rotating about an axis normal to the plane of the wall, as shown in Fig. 5.2a. In such a wall, known as a *Bloch wall*, the component of M normal to the wall is continuous, so that no volume magnetic poles are present. However, free magnetic poles are created at the intersection of a wall with the surface of the specimen, giving rise to magnetostatic energy. For a bulk specimen,

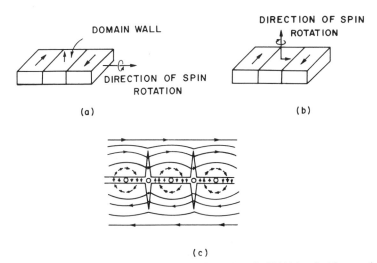

(a)   (b)

(c)

Figure 5.2. Spin configuration in domain walls: (a) Bloch wall; (b) Néel wall; (c) cross-tie wall.

since these poles are situated far apart, this component of energy is negligible, thus making Bloch walls energetically favorable in bulk specimens. For an in-plane magnetized film, below a certain thickness of the specimen the magnetostatic energy predominates over the other two components, making it unfavorable for the spins to deviate from the plane of the film, as is required for Bloch wall formation. In such cases, the spins within the wall take up the second configuration. In such a wall, known as a *Néel wall*, the spin changes from one direction to the other by rotating about an axis in the plane of the wall, as shown in Fig. 5.2b. In a Néel wall, poles at the surface of the film are eliminated, and instead volume poles are created within the wall, since now the component of M normal to the wall is not continuous.

Figure 5.3 shows the calculated[9] values of the wall energy and the wall thickness as a function of film thickness for a film with planar magnetization. It can be seen that (1) a Bloch wall is energetically more favorable than a Néel wall in bulk materials, and (2) a Néel wall is more favorable than a Bloch wall for films less than 350 Å thick. It has been found experimentally that as the film thickness decreases, a Bloch wall does not change abruptly into a Néel wall at a certain thickness. Instead, an intermediate structure known as the cross-tie wall is observed. A cross-tie wall

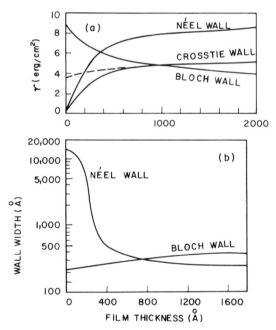

Figure 5.3. Theoretical variation of (a) surface energy density and (b) width of domain walls as a function of film thickness (after Ref. 9).

(Fig. 5.2c) may be thought of as a Néel wall having alternating intervals of oppositely directed senses of rotation of magnetization within the wall. This arrangement provides some flux closure in the wall.

As already mentioned, in a film having a negative perpendicular anisotropy constant (that is, $K_\perp < 0$) the magnetization vector lies normal to the film plane, and in a film having a net positive $K_\perp$, the magnetization vector lies in the plane of the film. For a film having a small value of negative $(K_\perp)_s$, $K_\perp$ changes sign from positive to negative above a certain critical thickness. However, the transition from in-plane to normal magnetization is not observed to be abrupt. In this intermediate thickness range a very interesting domain structure, called the *stripe-domain* structure, is observed. It consists of narrow stripe-shaped domains running parallel to each other and to the field previously applied in the plane of the film. When a planar field is applied perpendicular to the stripe walls and the intensity of the field is gradually increased, a reorientation of the stripe walls occurs and the stripes rotate irreversibly for fields higher than a critical value. A model of spin configuration has been proposed[10] according to which all the spin components in the plane of the film are parallel to the walls but the spin in each domain has an alternate sign of normal component. In films with EA normal to the film plane, cylindrical domains of reverse magnetization are produced when a magnetic field is applied along the EA. Such domains, also known as *magnetic bubbles*, are of great importance in computer technology and will be discussed in detail in Section 5.2.

5.1.4. *Switching in Thin Films*

Due to the presence of UA, the shape of the hysteresis loop of thin films, unlike that of ordinary bulk ferromagnetic materials, depends on the direction of the applied magnetizing field. Stoner–Wohlfarth's coherent rotation theory,[11] according to which magnetization of a film remains uniform at all times, can be applied for studying the switching behavior of an ideal single-domain film. Accordingly, the magnetization energy is

$$E = -M_s H_\parallel \cos \theta - M_s H_\perp \sin \theta + K_u \sin^2 \theta \qquad (5.3)$$

where $H_\parallel$ and $H_\perp$ are the components of the applied field along the EA and the HA, respectively, and θ is the angle which the magnetization vector M_s makes with the EA.

Differentiation of E and $\partial E/\partial\theta$ with respect to θ yields a rectangular hysteresis loop for fields applied along the EA, and a straight line with no hysteresis for fields applied along the HA, as shown in Fig. 5.4. One can see from Fig. 5.4a that there are two stable states of magnetization for

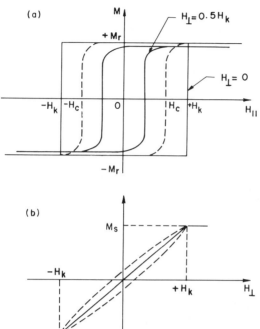

Figure 5.4. Hysteresis of a uniaxially anisotropic magnetic thin film (a) along the EA and (b) along the HA.

fields H applied along the EA, corresponding to two remanant states $+M_r$ and $-M_r$ of magnetization ($M_r = M_s$ in the ideal case). The critical field for a discontinuous irreversible change from one state to the other, i.e., switching, is $\pm H_k$, where $H_k = 2K_u/M_s$ is called the anisotropy field. For fields applied along the HA, magnetization varies linearly with field due to a gradual rotation of magnetization away from the EA, for field values up to H_k. For higher field values, the magnetization vector remains along the HA. Switching from EA to HA is a reversible process, as indicated by the absence of hysteresis in the M–H curve (Fig. 5.4b). We thus see that H_k is also the minimum value of the field required to pull magnetization from the EA to the HA.

There have been many studies of reversal processes in thin films using a variety of techniques.[12-15] These studies reveal that the reversal along EA does not take place by coherent rotation but by domain-wall motion in two steps: (1) growth of reverse domains along the EA, nucleated near the edges where the demagnetizing field is high, and (2) expansion of the elongated reverse domains by parallel shift in the hard direction, as shown in Fig. 5.5a. This also explains the fact that the experimentally observed coercivitiy H_c (i.e., the field at which half the magnetization has been reversed), also known as the domain wall coercive force, is less than the

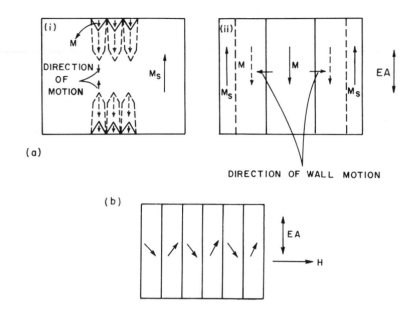

Figure 5.5. (a) Reversal by wall motion; (b) HA domain splitting.

theoretical value of $2K_u/M_s$ corresponding to reversal by coherent rotation, except for the special case of inverted films where $H_c > H_k$. The phenomenon of inversion is found in films having large magnetization dispersion. It has been implicitly assumed in the above discussion that the threshold field required for edge-domain-wall motion H_w (also called the wall-motion coercive force) is smaller than the field H_n for nucleation of a reverse domain. In the reverse case, which is not very common, it is expected that the EA magnetization reversal would take place by coherent rotation rather than by wall motion. The same is expected in the case of inverted films for EA field $H_\parallel$ ($H_k < H_\parallel < H_c$). However, the results of switching experiments indicate that due to dispersion in the EA, the reversal process is still predominantly one of domain-wall motion, preceded by incoherent rotation which divides the film into many domains with long axes parallel to the EA.

The theoretically predicted closed loop for fields $H_\perp$ applied along the HA exists only for values of $H_\perp \ll H_k$ in real films. At high drive fields the loop opens[16] up, as shown by the dashed curves in Fig. 5.4b. The finite area enclosed by the loop indicates the occurrence of some irreversible process during reversal. It has been found experimentally that when the field is gradually reduced after saturation along the HA, the film splits up into elongated domains with long axes parallel to the EA, as shown in

Fig. 5.5b. This phenomenon is called HA domain splitting or HA fallback, and is attributed to the presence of anisotropy dispersion in real films. It is the motion of these domain walls which is responsible for irreversible phenomena during HA reversal and hence the opening up of the HA loop. If a small EA field is applied when the HA field is removed after saturation, the film can be easily tipped in either sense of the EA depending on the direction of the applied EA field. This result is of great importance in computer-memory applications.

Let us now discuss the effect of simultaneous application of a dc bias field $H_\perp$ along the HA and a dc switching field $H_\parallel$ along the EA on the switching characteristics. Upon application of a field along the HA, the highly rectangular EA loop changes gradually to a collapsed S-shaped loop, as shown by the dashed curve in Fig. 5.4a for $H_\perp = 0.5H_k$. The effect of application of $H_\perp$ along the HA on the switching threshold H_s (i.e., the EA field at which rotational switching occurs) along the EA is shown[17] in Fig. 5.6a. One can see that the switching threshold continuously decreases from H_k at $H_\perp = 0$ to 0 at $H_\perp = H_k$. The decrease in H_s arises because, for a finite $H_\perp$, the equilibrium position of M would be away from the EA, so that a smaller field is required to overcome the energy maxima along the HA. Since the speed of switching increases with increase in field above the threshold, application of a bias field would decrease[18] the switching time τ for a given field $H_\parallel$ (see Fig. 5.6b). In other words, for a given τ, the field required for switching is correspondingly reduced. This is a very important conclusion for fast switching. It must be noted, however, that according to Fig. 5.4a the loop becomes less rectangular as $H_\perp$ increases further, which is not desirable for computer-memory applications.

The ranges of various switching modes are also shown in Fig. 5.6a. We can see that, in the region below the dc threshold curve, where the fields are too small for switching to take place even by domain-wall motion, switching can still take place by another process known as *creep*. Creep refers to the slow irreversible changes in magnetization when a small HA ac or pulse field is applied, along with a dc EA field well within the dc threshold curve. Creep is of great technological interest, as it means that the information stored in a film may be slowly lost under the repeated action of a pulsed field whose dc values would be too small to reverse the film.

We can also see that for fields just above the dc threshold curve, switching takes place by incoherent or inhomogeneous rotation (instead of coherent rotation) which involves labyrinth propagation[19] occurring due to local variations in the switching threshold. Here, reversal takes place by extension of the tip of the initial domain. A striking feature of labyrinth switching is that regions of unswitched material are left behind, resulting in a labyrinth-like flux pattern.

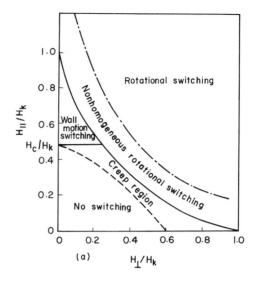

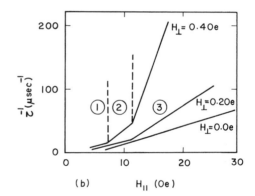

Figure 5.6. (a) Various modes of magnetization reversal in thin films (after Ref. 17); (b) reciprocal switching time τ vs. $H_{\parallel}$ for different values of $H_{\perp}$ (after Ref. 18).

Figure 5.6b shows that there are three distinct regions of switching depending on the values of the applied fields $H_{\parallel}$ and $H_{\perp}$. It has been directly verified[20] by monitoring the voltage induced in transverse and longitudinal pickup loops that (1) for $H_{\parallel}$ less than the threshold value H_s, where switching time is ~1 μsec (region 1), reversal takes place by domain wall motion, (2) for $H_{\parallel}$ just above the threshold value, where switching time is ~50–100 nsec (region 2), reversal takes place by incoherent rotation, and (3) for $H_{\parallel}$ much larger than the threshold value and with $H_{\perp} \neq 0$, where switching time is less than 50 nsec (region 3), reversal takes place by coherent rotation.

5.2. Applications

The various properties of magnetic thin films described in the previous section have been exploited for the fabrication of a large number of useful devices. These are described in the following sections.

5.2.1. Computer Memories

Magnetic films are attractive for this application because they can be produced with inherent bistable states (UA) and can be switched from one state to the other in a few nanoseconds under the action of small drive fields of the order of a few oersteds. Furthermore, large high-density arrays of film bits can be made in a favorable geometry at low cost.

5.2.1.1. Flat-Film Memory

A ferromagnetic film used as a memory element is typically an approximately 1000 Å thick film of Permalloy (81% Ni, 19% Fe) deposited on a smooth, flat substrate (such as glass) in the presence of a unidirectional magnetic field. The individual bit elements are fabricated either by deposition through suitable masks, or by selective etching of a continuous film.

A "1" ("0") is stored in a given bit when it is magnetized, say left (right) along the EA as shown in Fig. 5.7a. A memory must be designed to enable interrogation of each bit, so that the stored information may be read out, and provision must also be made to change the magnetization state at will, or to write in information. There are two common schemes for reading and writing of data: (1) the coincident-current scheme and (2) the word-organized or the cross-field scheme. In both these schemes, fields are applied in the plane of the film by driving current through conductor strips deposited on the film.

In the coincident-current scheme, widely used in ferrite-core memories, information at a given memory element, say I in Fig. 5.7b, is read by passing a suitable current through the conductors X_3 and Y_3, which gives rise to a magnetic field parallel to the EA. The value of the current in each conductor is such that the corresponding field value is $H_c/2$. In this case, element I would be subjected to a total field of H_c along the EA, half of which is supplied by X_3 and the other half by Y_3. On the other hand, elements, C, F, G, and H are each subjected to only a half field ($H_c/2$) while elements A, B, D, and E are not subjected to any field, as can be readily seen from an inspection of Fig. 5.7(b). As a result of this, only element I is capable of being switched. Hence, an output voltage would be detected in the sense loop oriented perpendicular to the EA, if the film were initially in the "1" state. If the film were in "0" state, no

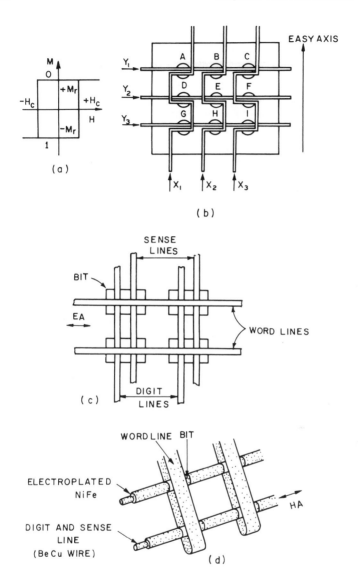

Figure 5.7. (a) EA M–H loop; (b) a nine-bit coincident-current matrix; (c) word-organized memory; and (d) plated-wire memory.

voltage would be detected in the sense loop. This process is therefore of "reading," as it allows us to detect whether a "1" or "0" was stored at I. For writing back "1" at I, which is necessary because the readout is destructive, a total field of $-H_c$ is applied to the element I.

It is clear that to produce a practical memory with coincident-current selection scheme, the coercive force H_c and the loop rectangularity must be the same for all memory elements to avoid switching of undesired elements. Also since the fields are applied along the EA, reversal takes place by the rather slow mechanism of domain-wall motion, which in the case of ferrite cores is the dominant reversal mechanism. In thin films, however, if the domain wall motion reversal is invoked, the high-speed potentialities of coherent or incoherent rotational modes remain unrealized.

High-speed coherent rotational switching is realized in the cross-field selection system shown in Fig. 5.7c. To interrogate a given memory element, say A, a current pulse is passed through the corresponding word line, thus creating a HA (or transverse) field pulse on all the elements in that row (word). The magnetization in each bit of the word will then uniformly rotate toward the HA, thereby generating a pulse in its associated sense line, of different polarity depending on whether a "1" or "0" is stored in that element. Upon termination of the word pulse, the information is destroyed by the HA domain-splitting process. The information must therefore be written into this element upon termination of the interrogating word pulse.

To write information, a pulse is again applied to the word line. Before termination of this pulse, another pulse is applied to the digit line, thereby generating an EA (or longitudinal) field in the element, which makes the magnetization relax back to the EA sense, corresponding to a "1" or "0" according to the sense of the digit pulse.

The main advantages of the cross-field scheme over the coincident-current scheme are: (1) a relaxation in the tolerances of the film parameters H_c and the rectangularity of the EA loop, (2) wide tolerances allowable in operating currents, and (3) ease in the amplification of the output signal in the presence of noise, since the readout signals corresponding to "1" and "0" are of equal magnitude with opposite phase. In the cross-field scheme, particular attention must be paid to the slow reversal of the unselected elements due to creeping. To reduce creep effects, specially shaped[21] elements and multilayer elements[22] have been suggested by some workers.

5.2.1.2. Cylindrical-Film or Plated-Wire Memory

Such a memory[23] typically consists of a parallel array of specially prepared Be–Cu wires onto which 81% Ni + 19% Fe alloy film has been electroplated, with a set of word lines placed perpendicular to the wires, as shown in Fig. 5.7d. The memory elements are defined by the intersections of the wires and the word lines and have circumferential easy axis due to the passage of current through the wires during electrodeposition.

The main advantage of this memory over the flat-film memory is that it provides flux closure in the elements thus reducing the demagnetizing fields. This leads to greater permissible film thicknesses and hence a larger readout signal. In addition, this geometry provides close coupling between the elements and the sense-digit wire. A disadvantage of the cylindrical-film memory is that the fringing fields from the word lines are larger than in the flat-film case, which necessitates bit densities at least an order of magnitude lower than those practical with flat films.

5.2.1.3. Magneto-Optical Memory

A magneto-optical memory differs from the previously described memories in that the data is written and read by a laser beam. Laser-beam addressing offers the following advantages: (1) since the bit size is only diffraction limited ($\sim 1~\mu$m), bit densities $>10^6/\text{cm}^2$ can be easily achieved; (2) since the readout signal is derived from the beam intensity, which is external to the memory element, a high signal-to-noise ratio is obtained; (3) since the optical beam is inertialess, much shorter access times are possible; (4) since the optical transducer is maintained at a distance from the storage medium, no head crash problems arise as in disk or drum memories; and (5) The NDRO (nondestructive readout) facility is available.

There are two methods of addressing with a laser beam: (1) the method in which the storage medium is stationary while the laser beam scans the entire memory plane, and (2) the method in which the storage medium rotates in its plane and the laser beam scans radially to address circular tracks, as an inductive magnetic head does in a conventional disk memory. Parallel information is also possible as in holographic memories (see Section 3.7). In magneto-optical memories information is written by thermo-magnetic recording which can be done in ferromagnetic materials by Curie-point writing and threshold writing and in ferrimagnetic materials by compensation-point writing. In Curie-point writing, the memory element is raised to temperatures above the Curie point where it loses its magnetization. When an external magnetic field is applied along the EA, the element is magnetized and the information is written. Curie-point writing has been demonstrated in thin films of Eu chalcogenides,[24] MnBi and $\text{Mn}_{0.8}\text{Ti}_{0.2}\text{Bi}$,[25] MnAlGe,[26] and MnGaGe.[27]

Threshold writing and compensation-point writing are based on the reduction of coercivity in the heated elements, enabling the magnetization to reverse under the influence of a field which is greater than H_c at elevated temperatures but less than H_c at the ambient temperatures. This has been demonstrated in EuO films,[28] MnAlGe films,[29] MnGaGe films,[30] Gd iron garnet, and amorphous Gd–Co films.[31,32]

The majority of the magnetic films used in magneto-optical memories are characterized by a UA perpendicular to the film plane (except Eu chalcogenide films). The most useful magneto-optic readout methods are based on the Kerr and Faraday effects, which involve rotation of the plane of polarization of the reflected and the transmitted beam, respectively. Since the rotation is sensitive to the sign and magnitude of the magnetization component along the beam direction, the two states "1" and "0" are placed in optical contrast when the film is viewed between crossed polarizers.

5.2.1.4. Ferroacoustic Memory

A ferroacoustic memory[33] is an all-electronic analog of drum and disk block-oriented mass memories having nonmechanical addressing, NDRO, high data-transfer rates, short access times ($\sim 1\ \mu$sec to a block), low power consumption, small size, and solid-state reliability.

A strain pulse is utilized to access an array of batch-fabricated thin film storage planes, called *modules*. Each memory plane consists of two UA thin films (each ~ 1000 Å) of magnetostrictive (Ni 60% + Fe 40%) alloy vacuum-evaporated on a glass substrate with a conductor layer of Cu (~ 5000 Å) between them and a top protective layer of SiO (~ 2500 Å). This multilayer system is etched into data lines as shown in Fig. 5.8.

Data is written when a magnetic field applied perpendicular to the propagation axis of the sonic pulse and generated by current pulses applied to the drive conductor is concident with the sonic pulse, which lowers the switching threshold of the material. Data is read out as a change in the remanant magnetization when a second sonic pulse propagates along the data lines. Since the data is read out serially as the strain propagates along the data lines, the same conductor can be used both as the drive and the sense elements. Information can be erased when a current pulse is initiated in the absence of a sonic pulse.

Since the data in this memory is read out without the application of a magnetic field, the signal-to-noise ratio is exceptionally high and is determined only by amplifier noise.

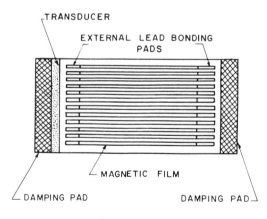

Figure 5.8. Top view of a ferro-acoustic memory plane.

5.2.2. Domain-Motion Devices

A large number of storage and logic devices based on the movement of domains in thin films under the influence of applied fields form an important subject of current investigations in the area of high-capacity digital storage, becuse they provide both high-density storage and high-speed logic in the same film. The combination of logic and memory within a single storage medium reduces the amount of associated drive and selection electronics. Also, these devices are easily adaptable to batch or automated fabrication techniques. These devices can be broadly divided into two groups: (1) domain-wall-motion devices, which involve domains magnetized in the plane of the film, and (2) magnetic-bubble devices, which involve domains magnetized normal to the film. In both cases, information is carried by domains of reverse magnetization in an otherwise saturated film, which are made to move through the storage medium (i.e., the film) by the application of suitable magnetic fields, in contrast to tape and disk systems where mechanical motion of the storage medium is involved, which limits the speed of data processing.

5.2.2.1. Domain-Wall-Motion Devices

The possibility of utilizing controlled motion and interaction of magnetic domains by wall motion for storing, propagating, and processing digital information has been examined by several workers. The first shift register based on a design proposed by Broadbent[34] involved domain propagation by in-phase translation of the walls to preserve the original configuration. This, however, was not given much thought because in spite of some improvements suggested by Tickle,[35] it was not possible to avoid the nucleation of some spurious domains due to the trailing edges of the moving domains during operation.

Another technique[36,37] of domain propagation which, in addition to lossless transfer of information, also provides the possibility of fan-out (or gain) of the output by expansion of the information-bearing domains, is based on domain tip propagation (DTP). DTP refers to the reversal of magnetization by growth of a reverse domain in the vicinity of the spike-like extremity of the domain, instead of sidewise expansion of the domain walls. DTP devices involve domain propagation in narrow channels of low-coercive-force (~2 Oe) NiFeCo film (1500 Å on glass) which are surrounded by high-coercive-force (~25 Oe) regions of NiFeCo. The difference in coercive force is achieved by depositing a rough layer of aluminum prior to NiFeCo deposition in regions where high coercivity is required. The propagation channels can be defined by the usual photo-lithographic technique. The magnetically hard regions surrounding the propagation channels not only restrict magnetization reversal to the channels but also play an important role in inhibiting the spontaneous formation of extraneous domains within the channels, by providing continuity of magnetization across the channel edges in the unswitched state.

The threshold field for DTP depends on a number of parameters such as the orientation of the channel with respect to the EA, channel width, film thickness, etc. Obviously, for controlled DTP, the value H_w of this threshold should be much smaller than the anisotropy field H_k to avoid the nucleation of reverse domains within the unswitched portion of the channel, which will lead to spurious signals. Addition of Co to NiFe increases H_k, so that fields much greater than H_w can be safely applied to obtain higher wall velocities and hence higher speeds of data processing.

Because of the directionality and high speed ($\sim 10^6$ cm/sec) of DTP and the possibility of strong magnetostatic interactions between closely spaced channels, this mode of domain growth is adaptable to a great variety of logical operations. Various types of shift registers of essentially unlimited length with high storage density and high-speed counters, buffer memories, push-down list memories, and logic elements based on DTP have already been fabricated.[38] The operation of one such shift register is described below.

The low-coercive-force channels form a zigzag pattern as shown in Fig. 5.9. Suppose a domain of reverse magnetization [boxed arrow in (a)] is introduced by a write conductor. Application of a field as shown in situation (b) in the figure makes the domain expand along the low-coercivity channel with magnetization direction nearest to the field direction. To restore the unswitched state of magnetization leaving a small reverse domain [boxed arrow in (c)], a field is applied as shown in situation (c), and its strength is reduced to values much lower than those required for reversal by the help of an inhibit conductor running beneath. Information is further shifted in the same way, as shown in situations (d) and (e). This type of shift register has been fabricated and tested by Spain[38] at data rates of 500 kHz with density typically 2000 bits/in.² and channel widths ~0.002 in. Prototype memories of up to 24 Mbit capacity have been built and operated satisfactorily.[39,40]

Attempts[41] have also been made to utilize the opposite spin configurations (Bloch and Néel) in a wall, to represent the binary information instead of the bistable state of a UA film.

5.2.2.2. Magnetic-Bubble Devices

Magnetic bubbles[42] are extremely small and highly mobile (under the influence of suitably applied magnetic fields) cylindrical domains of reverse magnetization, observed in thin films (or thin wafers) having an EA perpen-

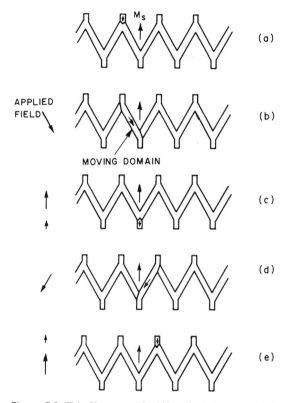

Figure 5.9. Thin film magnetic shift register (after Ref. 36).

dicular to the film plane. They were invented and first reported[43] by Bobeck in 1967 and were nicknamed "bubbles" because of their appearance when viewed under a microscope using polarized light. In order for the bubbles to move freely without any obstruction, the material must be boundary free, i.e., either a single crystal or amorphous. In order for a thin film to support bubbles, it must have a UA with EA perpendicular to its plane.

Let us now see how bubbles are formed. Figure 5.10a shows the domain pattern of a thin film with perpendicular EA in the absence of an applied field. The pattern consists of two regions of intertwined serpentine domains, one with upward and the other with downward magnetization. Upon application of a bias field normal to the plane of the film, the region with magnetization parallel to the field grows at the expense of the other with antiparallel magnetization (Fig. 5.10b). Eventually, the regions with magnetization opposite to the field shrink to circular cylindrical domains— the bubbles (Fig. 5.10c). Further increase of the bias field collapses these

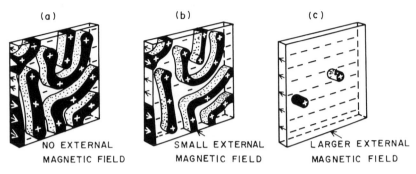

Figure 5.10. A schematic diagram showing formation of magnetic bubbles (after Ref. 51).

bubbles. However, on decreasing the bias field, the bubbles expand to restore the serpentine pattern.

As already discussed in Section 5.1.2, the condition for the existence of bubbles is a negative anisotropy constant of magnitude

$$|K_\perp| > 2\pi M_s^2$$

It has been shown by Thiele[44] that, under optimum conditions, the diameter d of a bubble and the thickness t of the film are related by

$$d \simeq 2t \qquad (5.4)$$

When Eq. (5.4) holds, the diameter of the bubbles is given by

$$d = 8(AK_\perp)^{1/2}/\pi M_s^2 \qquad (5.5)$$

The various materials which can support bubbles are summarized in Fig. 5.11, which was first drawn by Gianola *et al.*[45] and later updated by Chang.[46] Of these materials, orthoferrites and hexaferrites have perpendicular EA due to large magnetocrystalline anisotropy. It can be seen from the figure that in orthoferrites the bubbles are too large ($\sim 100~\mu$m) to be of use in high-packing-density applications. Moreover, these materials are very difficult to prepare in thin film form because of lack of suitable substrates for epitaxy. On the other hand, although in hexaferrites the bubble size is small, the bubble mobilites are too low[46] for high-speed applications. The most important materials for bubble-device fabrication are garnets and amorphous alloys. Although garnets are crystallographically cubic, slight deviations from cubicity during growth are sufficient to induce the required UA. Garnet films are grown by using sputtering, CVD, and LPE techniques.

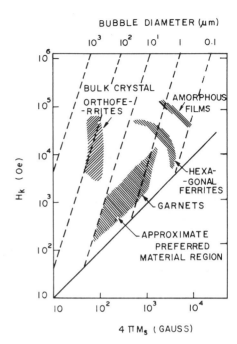

Figure 5.11. Magnetic-bubble materials status chart (after Refs. 45 and 46).

Magnetic bubbles in amorphous alloy films were first reported by Chaudhari *et al.*[47] in 1973 for sputtered Gd–Co and Gd–Fe films. These amorphous alloys may be ferromagnetic or ferrimagnetic and differ from magnetic insulators and transition metal alloys in that the exchange mechanism leading to magnetic ordering is an indirect one involving conduction electrons. The UA in these structurally homogeneous and isotropic materials is due to short-range pair ordering. Other magnetic-bubble amorphous materials which have been studied include GdCoAu,[48] GdCoCu,[48] GdCoMo,[49] GdCoCr,[49] and GdCoNi.[50] Of these GdCoMo has properties closest to those of garnets. Certain features of bubbles which are noteworthy for device applications are: (1) stable bubbles exist over a range of bias fields providing stable storage, (2) a bubble can be elongated by lowering the field for further manipulation, for example, replication for generation of bubbles and for NDRO, (3) a bubble can be annihilated by raising the bias field for storage clearance, and (4) bubbles can be propagated without spurious bubble generation because the nucleation fields are much greater than the wall-motion fields.

We now describe how bubbles can be propagated and controlled[51,42] for storing data and performing logical operations. This requires the creation of in-plane magnetic driving fields and can be achieved using two methods. The first, the conductor-access method, employs an array of

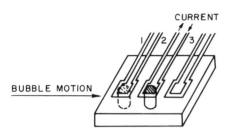

Figure 5.1.2. A simple conductor arrangement for moving bubbles.

conductors placed on the film by lithographic techniques, as shown in Fig. 5.12. The size of the conductor loops is of the order of the bubble diameter. Fields are produced by passing current through these conductors. The conductor which is energized attracts the bubble into its loop. The problem of interconnecting a large number of conductors with external access circuits is greatly reduced in the second method, called the field-access method.

In one such method, the overall magnetic bias on the film is rythmically raised and lowered, so that bubbles alternately shrink and expand. Propagation of bubbles results if a pattern of closely spaced asymmetrical energy traps, e.g., permalloy arrowheads (affectionately known as angel-fish), is deposited on the film as shown in Fig. 5.13a.

The T-bar bubble mover is another field-access scheme which is more promising than the previous one. Here the bubbles remain the same size and are propagated by a rotating magnetic field in conjunction with a pattern of thin film permalloy T's and vertical bars. Figure 5.13b clearly shows the movement of bubbles using this scheme.

Each of these access methods can be used to generate new bubbles. In the conductor-access method, a loop conductor equipped with a hairpin conductor which can be separately energized acts as a bubble generator. When the hairpin conductor is energized at the instant the bubble lies on it, the bubble is pinched at the waist and fissions into two in less than a few millionths of a second. In the T-bar scheme, the generator is a permalloy

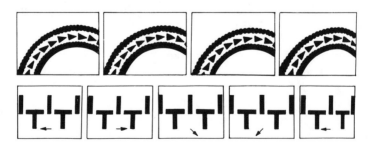

Figure 5.1.3. Angel-fish bubble mover and T-bar bubble mover (after Ref. 51).

disk with a small horizontal projection. A bubble can be removed by an annihilator or a bubble eater, which is simply a region of high bias field. Bubbles can be detected using either direct optical sensing or magnetoresistance sensing.

A large number of storage and logic devices have been proposed[46] using magnetic bubbles. Very recently a 1-Mbit bubble memory device (RMB 411), fabricated using 2 μm technology, has been commercialized by Rockwell International. It is composed of 572 storage loops, each containing 2052 bubble positions. This device is compatible with the Rockwell AIM 65 microcomputer. The memory is block oriented with four modules each of (6 × 9.75) in.^{2} size and operates at 100 kHz field rate.

5.2.3. Thin Film Magnetic Heads

Magnetic heads are transducers which convert electrical signals into a corresponding magnetization pattern on a recording medium and vice versa. The former type are called *recording* (or write) heads and the latter type are called *replay* (or readout) heads. Although the performance of conventional magnetic heads is satisfactory for a large number of applications, some of the present and future high-density recording applications essentially dictate the development of miniature integrated thin film heads. Thin film heads, being two-dimensional, have the advantage of simple installation and alignment, small size, light weight, and feather-touch contact with the recording medium. The most important type of heads are based on the principle of (1) electromagnetic induction, (2) magnetoresistance, and (3) the Hall effect.

5.2.3.1. Inductive Heads

Inductive heads[52] are based on Faraday's law of electromagnetic induction and can be used in both write as well as read modes. The simplest kind of inductive head is a single-turn head, the two possible geometries of which are shown in Fig. 5.14a. It is a solid-state analog of the conventional head, which consists of a ring-shaped magnet with a gap in it and a coil wound around it as shown in Fig. 5.14b. Longitudinal magnetic recording (i.e., magnetization direction parallel to the direction of motion of the recording medium) takes place on a recording medium in the region under the gap by the fringing field across the gap produced by the circumferential flux when a current corresponding to the signal is passed through the coil. During reproduction of the original signal, the magnetic field of the recorded information under the gap induces a voltage in the coil. Romanikiw *et al.*[53] have been able to batch-fabricate such single-turn heads with gap

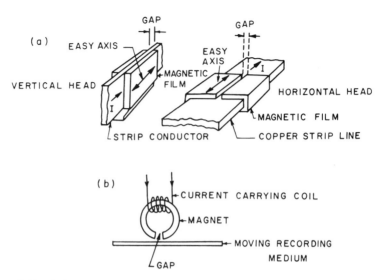

Figure 5.14. (a) A single-turn thin film recording head (after Ref. 52); (b) a conventional recording head.

~2 μm and magnetic film (permalloy) thickness ~2.5 μm using electro-polating, photoetching, and electron beam lithographic techniques. The EA of the magnetic film is made parallel to the length of the gap so that the film is driven in the hard direction. This is done to improve the frequency response of the device, because when the film is driven in the hard direction, magnetization changes take place rapidly by rotation rather than by the relatively slow process of domain-wall motion. The horizontal geometry has the advantage that it can be easily fabricated using lithography techniques, but it is more vulnerable to wear and tear problem due to abrasion from the recording medium. On the other hand, the vertical geometry is more stable but requires techniques like micron scale lapping and grinding for fabrication.

Another slightly different version of this head which also provides amplification of the output during readout is due to Kaske *et al.*[54] It differs from the previous one in that (1) the EA of the magnetic film (~2000 Å thick) is perpendicular to the length of the gap and (2) it has an additional conductor (read line) at right angles to the previous one (sense/drive-line) placed on the side opposite to the gap and separated from the magnetic film by an insulating layer of SiO. Longitudinal recording is achieved by using the sense line as the drive line. Recording with densities as high as ~1500 flux reversals/cm have been demonstrated in a 230 Å thick film of Fe with H_c ~ 80 Oe. The readout is accomplished by passing a current through the read line which saturates the film along the HA. On reducing this HA field, the field from the recording medium steers the head magnetization from HA to EA (the polarity of which is determined by the recorded information), thus producing a voltage in the sense line. Amplification is produced because the sensed flux is produced from the head magnetization and not from the recorded magnetization, which only acts as a steering field. For the same reason, the readout is independent of the speed of the recording medium. Also since a very small steering field is required, recording can be performed even in thin media, which have the advantage of being capable of sustaining high recording densities even at low coercivities.

Multiturn heads are very attractive because they give much higher outputs at much smaller currents as compared to single-turn heads, but their fabrication is much more complicated. A number of designs have been proposed[55,56] for multiturn heads. A multiturn coil is made by depositing alternate layers of a conductor (say Cu) and an insulator (SiO_2). The gap is fabricated either by evaporation through masks or by lithographic techniques. The operation of a multiturn head is similar to that of a single-turn head. Hanazono *et al.*[57] have developed an eight-turn, 27-track arrayed head (specifications: track width 170 μm, gap width 1 μm, and track pitch 940 μm) using magnetic layers of vacuum-evaporated Ni-Fe (3 μm), conductor layers of sputtered Mo–Au–Mo multilayered structure, and insulator layers of sputtered SiO_2 on a Si substrate. The gap was fabricated by photolithography.

5.2.3.2. Magnetoresistive Heads

The resistivity of a magnetoresistive material[58] in the presence of a magnetic field is given by

$$\rho = \rho_0 + \Delta\rho \cos^2 \theta \qquad (5.6)$$

where ρ_0 is the isotropic resistivity in the absence of the field, $\Delta\rho$ is the magnetoresistance coefficient, and θ is the angle between the magnetization vector and the current-density vector. Basically a magnetoresistive head consists of a thin ($\sim$1000 Å) film of permalloy, the $\Delta\rho$ for which is of the order of 2–6% of ρ_0, placed with its plane either parallel (horizontal configuration) or perpendicular (vertical configuration) to the storage medium. A magnetoresistive head has the limitation that it can be used only in the read mode. The fringing field of the stored information changes the direction of magnetization of the film and hence the resistance of the film [Eq. (5.6)]. This change in resistance can be detected in a number of ways. One way is to supply a constant current to the device so that an output voltage proportional to $\cos^2 \theta$ appears. As in the case of inductive heads, the vertical geometry is preferred from the point of view of wear resistance.

5.2.3.3. Hall Heads

Hall heads[59] are based on the principle of the Hall effect which states that when a current-carrying conductor is placed in a transverse magnetic filed, an emf is generated in a direction perpendicular to both the current and the field directions. Thin film Hall heads have been fabricated[60] by Hitachi Ltd. for use in stereo tape recorders, using low-noise, 1.4-μm-thick vacuum-evaporated InSb Hall elements. The Hall heads are readout heads and have the following advantages over the inductive readout heads: (1) the output voltages are much higher, (2) the output voltage is independent of the signal frequency and the tape speed because the Hall heads detect directly the magnetic flux whereas the inductive heads detect the time derivative of the flux, (3) the output can be increased by increasing the dc bias current, and (4) signal-to-noise ratio is more than 10 dB higher in the low-frequency range, and 2–4 dB higher throughout the audio range. These advantages result in hi-fi (high-fidelity) reproduction of audio tape sounds. Other applications of Hall heads, such as multichannel pickup of video signals and digital data, are also very promising.

5.2.4. Magnetic Displays

Magnetic displays are not commonly used because of the availability of a large number of other more useful displays described in Chapter 3. Nevertheless, for the sake of scientific interest we would like to mention here briefly the underlying principles.

One such display[61] utilizes stripe domains in Ni–Fe films. Due to the field from the domains, the overlying suspension of colloidal iron oxide particles on the film is drawn into long parallel lines acting as an optical diffraction grating which can be altered by the applied magnetic field. When the display panel is illuminated with an intense source of light, the elements which are magnetically turned "on" reflect light and look bright (with color and intensity depending on the applied filed), while the ones that are turned "off" remain dark.

Another type of solid-state magnetic display utilizes[62] thin magnetic films in conjunction with insulator and conductor layers. Two geometries, as shown in Fig. 5.15a and b, can be used. Each consists of small regions of low coercivity separated by regions of high coercivity. In geometry (a), regions of permalloy film deposited on to Al have high H_c (~110 Oe) and those deposited on SiO have low H_c (~4.5 Oe). In geometry (b), which utilizes two coupled films, the regions which are coupled through SiO have

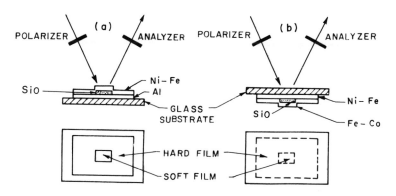

Figure 5.15. Schematic diagrams of magnetic display elements.

low H_c and the ones coupled directly have high H_c. The display pattern is determined by the state of magnetization in the low-coercivity regions, which may be changed by the pulse currents flowing through the selected elements. The display pattern can be observed by the Kerr effect when the panel is illuminated with plane-polarized light.

6

Quantum Engineering Applications

6.1. Introduction

The discovery of superconductivity by Kammerlingh Onnes[1] in 1911 and of the associated effects by others in subsequent years has given birth to a new branch of engineering called quantum engineering. Studies of superconductivity in thin films have made remarkable contributions to the development of the physics of this phenomenon, which is now well established as a macroscopic manifestation of quantum effects. One of the most fascinating phenomena revealed by thin film studies is that of superconductive tunneling, which alone has led to the emergence of a completely new technology of superconducting electronics. Superconducting integrated circuits make the most accurate electronic devices. This chapter discusses the whole variety of thin film superconducting devices. It is not our aim here to discuss the physics of superconductivity, exhaustive details of which can be found[2-6] in standard textbooks and reviews available at all levels. Nevertheless, to maintain the continuity of ideas, before going over to the discussion of devices we would like to give a brief account of some of the fundamental concepts of superconductivity which are pertinent to devices.

6.2. Basic Concepts

Superconductivity is the phenomenon of the abrupt transition of a material from a normally conducting state to a state of zero dc electrical resistivity below a critical temperature T_c in zero magnetic field. The width of this transition can be as low as $\leqslant 10^{-4}$ K for pure and structurally perfect elemental specimens but may increase by more than two orders of magnitude for some alloys and compounds. The superconducting state has a lower entropy than the normal state, suggesting that the conduction electrons below the transition temperature are in a highly ordered state. The nature of this ordering will be discussed a little later.

Another dramatic property of the superconducting state is the exclusion of magnetic flux (Meissner effect[7]) and hence electrical currents from the bulk of the superconductor, both being largely confined to a surface layer ($\sim 10^{-5}$ cm thick) called the penetration depth, λ, which is a characteristic of the material and varies with absolute temperature T as

$$\lambda(T) = \lambda(0)[1 - (T/T_c)^4]^{-1/2} \tag{6.1}$$

where $\lambda(0)$ is the penetration depth at absolute zero. Flux exclusion takes place irrespective of whether the magnetic field is applied before or after the resistive transition. Actually, it is the field due to the screening supercurrent flowing in the surface layer which exactly cancels the applied magnetic field inside the superconductor, thus resulting in flux exclusion. Below the critical temperature, the superconducting state can be destroyed by the application of a magnetic field greater than a critical value H_c, the temperature dependence of which is given by

$$H_c(T) = H_c(0)[1 - (T/T_c)^2] \tag{6.2}$$

It should be noted here that the existence of a critical magnetic field also implies the existence of a critical transport current which can flow in a superconducting wire; this corresponds to the value of the current which produces a magnetic field H_c at the surface of the superconductor (Silsbee hypothesis[8]). With respect to flux penetration above a critical field, two distinct types of behavior are observed. In type I superconductors (elemental except Nb and V), flux penetration takes place only for applied fields greater than H_c. However, the flux penetration is uniform and complete only if the demagnetization factor of the sample is zero and thus the field at the superconductor surface is uniform. Due to finite demagnetization effects, the applied field in the vicinity of the superconductor is not uniform and may exceed H_c at some points where flux penetration starts, while remaining below H_c at others where there is no flux penetration. A type I superconductor containing such alternating superconducting and normal lamellae is said to be in the *intermediate state*. With increasing applied field, the normal lamellae grow at the expense of the superconducting ones until finally the specimen becomes completely normal and flux penetrates everywhere.

For a type II superconductor (Nb, V, and alloy superconductors), on the other hand, two critical fields H_{c1} (lower) and H_{c2} (upper) exist. For applied fields less than H_{c1}, the behavior of a type II superconductor is the same as that of type I, that is, it shows both zero resistivity and the

Meissner effect. At fields just above H_{c1}, however, the flux starts penetrating the specimen in microscopic tubular filaments (diameter $\sim 10^{-5}$ cm) called fluxoids or vortices lying parallel to the applied field, although the electrical resistance remains zero. Each fluxoid consists of a normal core with the total magnetic flux exactly equal to a fundamental quantum of magnetic flux ($\Phi_0 = h/2e = 2.07 \times 10^{-7}$ G cm^2 = 2.07×10^{-15} Wb) and surrounded by a superconducting sheath where a vortex of persistent supercurrent flows, maintaining the flux in the normal core. The current circulating around each core has a sense of rotation opposite to that of the surface screening current which keeps the flux excluded from the bulk. The fluxoids tend to arrange themselves in a regular lattice if the superconductor is sufficiently pure and defect-free. As the field is increased beyond H_{c1}, the fluxoids tend to grow in number until at field H_{c2} the superconductor is driven normal and the flux penetrates everywhere. (In some cases, however, a thin superconducting sheath at the surface may persist up to an even higher field $H_{c3} \approx 1.7 H_{c2}$). For applied fields between H_{c1} and H_{c2}, the superconductor is said to be in a *mixed state*. In contrast to type I superconductors where $H_c \leqslant 1000$ Oe, the value of H_c for type II superconductors may approach several hundred thousand oersteds. Since the electrical resistance in the mixed state is zero, a type II superconductor in this state can carry lossless currents even in the presence of very large magnetic fields. However, a small power dissipation does take place because of the movement of fluxoids due to flow of current. This power loss can be reduced by introducing defects into the crystal structure of the superconductor, which tend to pin down the fluxoids thus preventing them from moving.

Another important experimentally observed property of a superconducting ring (or a hollow cylinder), which was first predicted by London[9] in 1950, is that the flux contained in the hole of a ring due to a circulating persistent current is quantized in units of h/q (where q was later shown to be equal to $2e$ instead of e, the electronic charge, as predicted by London) in contrast to being arbitrary, as for the case of a ring in the normal state. Yet another characteristic property of a superconductor which distinguishes it from a normal metal conductor is the existence of an energy gap ($\sim$millielectron volts) between the ground state and the lowest excited state. The presence of an energy gap in the excitation spectrum has been confirmed by a variety of experimental measurements such as those of electron heat capacity, thermal conductivity, far IR, and microwave absorption and tunneling.

The various theories of superconductivity proposed to explain the above experimental observations fall into two categories: phenomenological and microscopic. A phenomenological treatment was first initiated by the brothers F. London and H. London,[10] who modified Maxwell's electromagnetic equations to allow for the Meissner effect and predicted the existence and order of magnitude of the penetration depth. Further treatment of the phenomenological theory by Ginzburg and Landau[11] and by Pippard[12] particularly emphasized the concept of the range of coherence. According to Pippard, the dependence of supercurrent on vector potential is nonlocal, that is, the supercurrent at a point is dependent on the vector potential throughout a region of size ξ (called the coherence length) instead of simply the vector potential at that point as assumed by the London brothers. Ginzburg and Landau (G–L) introduced an order parameter Ψ (similar to the quantum-mechanical wave function) to describe the "strength" of the superconducting state. The magnitude of the order parameter is proportional to the fraction of electrons in the ordered superconducting state and is zero for the normal state. Ginzburg–Landau equations relating the supercurrent to the order parameter are intrinsically nonlocal and naturally lead to the existence of penetration depth λ and coherence length ξ. Here ξ represents the distance over which the order parameter changes from a finite value to zero at the boundary between a superconducting and a normal region. Based on the Ginzburg–Landau theory, Abrikosov predicted[13] the behavior of a type II superconductor. He also stated that a superconductor behaves as type I if the Ginzburg–Landau parameter κ ($\approx \lambda/\xi$) is less than $1/\sqrt{2}$, and as type II if $\kappa > 1/\sqrt{2}$.

It was London[9] who first described superconductivity as a macroscopic quantum-mechanical state with a wave function $\Psi = \Psi_0 \exp(i\Phi)$ (where Ψ_0, the amplitude of the wave function, is constant for a uniform superconductor and Φ, the phase factor, is in general a function of space and time and also involves the magnetic vector potential), and who predicted flux quantization of a superconducting ring on the basis of the single-valuedness of the wave function at a given point.

A microscopic theory was proposed by Fröhlich[14] who was the first to realize the importance of electron–lattice interaction in explaining superconductivity, the experimental verification of which followed from the observation of the isotope effect (i.e., the proportionality of T_c and H_c to $M^{-1/2}$, where M is the isotope mass). A detailed microscopic theory, the BCS theory as it is called, was developed by Bardeen, Cooper, and Schrieffer[15] in 1957 and represents one of the most outstanding landmarks in the theory of solids. According to this theory, superconductivity is due to the condensation of the conduction electrons into a lower energy state in which electrons of opposite spins and equal and opposite momenta are paired (Cooper pairs) due to an attractive, phonon-mediated interaction. This lower-energy ground state is separated from the normal state by an energy gap 2Δ (centered around the Fermi level) which decreases from $3.52kT_c$ (k is the Boltzmann constant) at $T = 0$ to zero at T_c and which can be regarded as the binding energy of each of these Cooper pairs ($\uparrow \mathbf{k}$, $-\mathbf{k}\downarrow$). The energy Δ required to produce one excitation (also called quasiparticle or normal electron) is called the gap parameter. Because of the presence of the energy gap, the resistivity at $T = 0$ remains ideally zero up to a critical frequency $\omega_c = 3.52kT_c/\hbar$ after which it attains a finite value which increases with increasing frequency due to the quasiparticles produced by breaking up of the Cooper pairs. At finite temperatures ($<T_c$), however, there is a finite ac resistivity for all $\omega > 0$ due to thermally produced quasiparticles. For $\omega \gg \omega_c$, the resistivities of the superconducting and the normal state are essentially equal, irrespective of the temperature. The size of a Cooper pair ($\xi = hV_F/\Delta$, where h is the Planck's constant and V_F is the electron Fermi velocity), which corresponds to the coherence length, is typically $\sim 10^{-4}$ cm. This means that any given Cooper pair will overlap about 10^6 other pairs, showing that momentum ordering takes place not only within a pair but also among different pairs. A superconducting state is thus characterized by its ground state wave function $\Psi(r, t) = \Psi_0 \exp(i\Phi)$, where Ψ_0 is the amplitude and Φ is the phase factor of the wave function. The BCS ground-state wave function corresponds to the order parameter in the Ginzburg–Landau theory. The BCS theory accounts for all the unique properties of a superconductor, and the results of the various phenomenological theories can be derived from it.

6.3. *Superconductivity in Thin Films*

The existence of a penetration depth (λ) and a coherence length (ξ) suggests that the superconducting behavior of specimens with at least one dimension of the order of λ or ξ, such as thin films, should be different from that of the bulk, and it is indeed so. In addition to this dimensional effect, the behavior of thin films may also differ from that of the bulk due to the presence of certain inherent features, such as stresses, impurities, imperfections, and smaller grain size, which are generally introduced during the fabrication process of thin films depending on the deposition parameters.

The change in T_c of a thin film takes place in accordance with the nature of the stress present. For example, tensile stress increases, whereas compressive stress decreases, the T_c of Sn films,[16] resulting in T_c values anywhere between 3.53 to 4.15 K (bulk $T_c = 3.73$ K). Although changes in T_c are small, the presence of impurity and stress inhomogeneities may produce substantial broadening of the transition.

Anomalous increases in value of T_c have also been observed in fine-grained amorphouslike films of a variety of materials. For example, Bi and Be, which are not superconducting in bulk down to 10^{-2} K, are[17,18] superconducting with $T_c = 6$ and 8 K, respectively, in amorphous vapor-quenched films obtained by vacuum evaporation onto substrates held at

Table 6.1. Superconductivity Enhancement (Ratio of the Transition Temperature of the Film T_{cF} to That of the Bulk T_{cB}) and Other Characteristics of Several Materials (After Ref. 6)

Material	T_{cF}/T_{cB}	Thickness (Å)	Grain size (Å)	Deposition technique[a]
Al/SiO/Al (multilayer)	4	60		VQ
Al	2.27			VQ
Mo	~7	200–4000	50–1000	IBS
W	~400	100–4000	<50	S, EBE
	~600	200–4000	50–1000	IBS
Zn	1.55			VQ
Sn + 10% Cu	~2		~40	VQ
Be	>800			VQ
Bi	>600			VQ
Ga	6.5	>2000		VQ

[a] VQ, vapor quenched by deposition at 4.2 K; EBE, electron-beam evaporation; S, sputtering; IBS, ion-beam sputtering.

liquid-He temperature. This phenomenon of substantial increase in T_c is called *enhanced superconductivity*. Superconducting enhancement values for thin films of some of the materials are shown in Table 6.1.

Another effect observed in thin films and of great importance in devices is the depression of T_c of a superconducting film by an overlayer of a normal metal (proximity effect). The effect is more pronounced if the normal metal is ferromagnetic because of the additional interaction of spins with the conduction electrons. A superimposed magnetic metal film may also give rise to what is known as gapless superconductivity.

The behavior of superconducting thin films in magnetic fields is extremely complicated[5] and depends on a number of parameters such as the magnitude and direction of the applied field, temperature, penetration depth, coherence length, mean free path, film thickness, and even the width of the film. However, it suffices here to mention only that because the film thickness is comparable to the penetration depth, the effective diamagnetism (i.e., flux exclusion) of a thin film is far from perfect. It can be shown[4] on the basis of thermodynamic calculations that this results in the critical field values being higher, approximately in the ratio $\sqrt{24}\,\lambda/d$ (d is the film thickness), than the bulk value. It has also been shown by Tinkham[19] that in a transverse magnetic field, thin films of all materials behave as type II superconductors.

For the commonly used geometry of a flat rectangular film, the current is nonuniformly distributed over the surface (being maximum along the edges) in an attempt to minimize the self-field component perpendicular to the film. As a result of this, the critical current of a thin film is reduced by 20–100 times than what it would have been if the current density were uniform. Fortunately, it is possible to achieve uniform current density in either of two ways: (1) by depositing the film on a cylindrical substrate of radius much greater than the film thickness and of length much greater than the radius, and (2) by depositing the film on an insulated ground plane, also called the shield plane, consisting of an extended superconducting layer of thickness much greater than the penetration depth and with T_c and H_c higher than the deposited film. In the latter case, an image current of polarity opposite to that of the film current flows in the ground plans, resulting in the net field being parallel to the film surface. Consequently, the optimum current distribution becomes uniform and therefore the critical current values are appreciably increased. Further, the inductance of the film-shielded plane combination is about two orders of magnitude smaller than the isolated film. This also implies that the mutual inductance between the film and the circuit elements (which are not deposited directly above the film) is about two orders of magnitude lower. These properties of the film-shielded plane combination make it very useful in certain devices.

Our discussion of devices is classified into three main sections. In the first section are described those devices which utilize the superconducting-to-normal (S–N) transition of a superconductor. Devices based on the phenomenon of superconductive tunneling are dealt with in the second section. The remaining miscellaneous applications are grouped together in the last section.

6.4. *S–N Transition Devices*

A superconductor can be switched from the zero-resistance supercon-ducting state to the finite-resistance normal state by the application of a magnetic field, an electrical current, or heat. This property of a supercon-ductor can be utilized for switching as well as for steering into alternative paths of electronic signals. Switching of ac signals (or dc signals for zero-resistance loads) can also be accomplished by changing the self-inductance or mutual inductance of the circuit elements by switching a nearby supercon-ducting shield between the S and N states. The various devices based on the above principle are described below.

6.4.1. *Switching Devices*

The magnetic field-induced S–N transition has been the most widely used effect for switching. Such a four-terminal current switch, in which the impedance between the two output terminals is controlled magnetically by means of an electrical current passed through the two input terminals, is called a *cryotron*. The earliest form of this switch was the wire-wound cryotron (WWC), consisting of a control coil wound around a central gate wire, as shown in Fig. 6.1a. In operation, the gate is switched from the superconducting to the normal state by the magnetic field generated by passing current through the control coil. The control coil is made of a

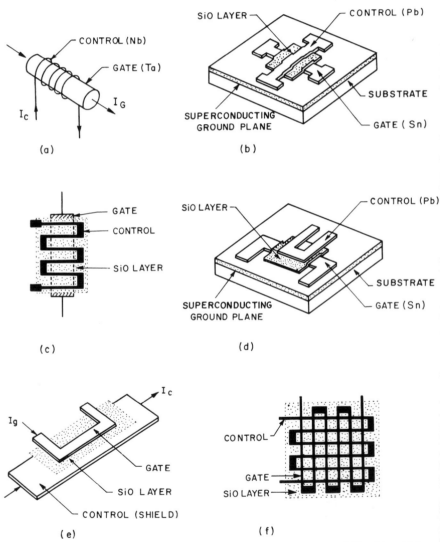

Figure 6.1. Several configurations of cryotrons: (a) wire-wound; (b) crossed-film; (c) multiple-crossing control; (d) in-line; (e) Ryotron; (f) multi-crossover cryotron amplifier.

material having H_c and T_c much higher than that of the gate wire, for example, Ta gate with Nb control, or Sn gate with Pb control, so that the control remains superconducting throughout the operation, which is carried out at temperatures slightly below the T_c of the gate and far below that of the control material. The switching time of the wire-wound cryotron (L/R) is severely limited by the large inductance L of the control winding and the small normal-state resistance R of the gate wire and is[20] typically ~50–200 μ sec. A significant increase (by three to five orders of magnitude) in switching speed can be obtained by using two-dimensional thin film equivalents of the wire-wound cryotron.

One of the common forms of thin film cryotrons is the crossed-film cryotron (CFC), shown in Fig. 6.1b. It consists of a gate film, usually of Sn, crossed by a much narrower control film and Pb, the two being electrically isolated by a thin film of SiO. The performance of the CFC is considerably improved by depositing the entire structure on a super-conducting ground or shield plane. The decrease in self-inductance of all the conductors situated near the shield plane (see Section 6.3) in this configuration leads to higher switching speeds and allows high component density due to lower pickup from neighboring devices. In operation, the gate is held just below its T_c as in the WWC and is current biased. The variation of the gate resistance with the control current at constant gate current exhibits a critical threshold value of the control current above which the gate resistance appears. The static gain of a shielded CFC at a given temperature is defined as

$$G = I_{gc}/I_{cc} \qquad (6.3)$$

where I_{gc} is the critical gate current in zero magnetic field (i.e., at zero control current) and I_{cc} is the minimum control current required to drive the gate resistance for very low gate currents. Expressing the currents in terms of the corresponding fields, we obtain

$$G = (H_{gc}/H_{cc})(W_g/W_c) \qquad (6.4)$$

where W_g and W_c are the widths of the gate and the control, respectively, and $H_{gc}/H_{cc} \sim 1$ for films much thicker than the gate penetration depth. Thus, by making $W_g \gg W_c$, current gains much greater than unity can be obtained. The gate current of such a cryotron can therefore be used to drive the control of another cryotron amplifier. Switching times as small as 0.4 μsec can be obtained[21] for a shielded CFC. A cryotron is very attractive because of its infinite on–off resistance ratio, but its on or open-gate (i.e., normal state) resistance is low, resulting in low output voltages which in some cases might not be able to drive the generally noisier room-temperature electronic devices. Higher values of output resistance can be obtained by using a multiple-crossing control film (Fig. 6.1c) instead of the simpler one of CFC, so that a larger portion of the gate is driven normal. It should be noted, however, that when the gate of this cryotron drives the control of another similar one, there is no decrease in the time constant because an increase in the gate film resistance is accompanied by a corresponding increase in the control inductance.

Another common geometry of a thin film cryotron is the in-line geometry shown in Fig. 6.1d. In the in-line cryotron, the control is also influenced by the gate current. This results in smaller I_{cc} when the two currents are parallel than when they are antiparallel, and therefore an asymmetry in the electrical characteristics. The static gain of the in-line cryotron is less than unity. However, a dynamic gain greater than unity can be obtained by biasing the gate to the steepest part of the asymmetrical characteristics. The switching time of the in-line cryotron[22] is $\sim 10^{-8}$ sec.

Cryotrons are ideal switching devices because of their small size, low power consumption ($\sim$microwatts, three orders of magnitude smaller than that of other conventional high-speed devices), and ease of fabrication and interconnection. Further, they are without input–output interaction, are level and polarity independent, are suitable for storage and logic, and require no auxiliary components such as resistors, diodes, transformers, or capacitors.

In addition to the resistive switching described above, electrical signals may also be switched by changing the inductance of circuit elements, for example, by switching a nearby superconducting shield between the superconducting and the normal state. One common type of thin film inductance switch is the ryotron (Fig. 6.1e), which consists of a narrow superconducting gate in close proximity to a wide shield conductor and separated by an insulating layer of SiO. This type of switch utilizes current-induced S–N transition. The shield is switched from the S to the N state by passing a current through it, as a consequence of which the inductance of the gate film changes from a low value to a high value while it remains in the superconducting state. Inductance switches are primarily used for switching purely superconductive loads such as cryotron control lines or superconducting memory drive lines.

Not much use has been made of thermally induced S–N transition in signal switches. However, thermally controlled switching is almost exclusively used in persistent current switches in high-critical-field magnets, where cryotrons are obviously not suitable. Also, for switching of elements in a sensitive magnetometer circuit, one has to resort to thermal means because the magnetic field used in cryotrons may induce noise in the circuit.

So far we have been discussing electrical switches. Since the thermal conductivities also differ widely in the S and N state, similar devices may also be used as thermal valves for controlling heat flow. To open the valve for allowing heat flow, the superconductor is switched from the low-thermal-conductivity S state to the high-thermal-conductivity N state. Such thermal valves are commercially used in very-low-temperature (<1 K) refrigerators.

6.4.2. Cryotron Amplifiers

Similar to other conventional switches such as triodes and transistors, cryotron switches can also be used in amplifying circuits. The design of high-gain multistage amplifiers using cascaded cryotron switches requires optimization of a number of parameters such as thickness and width of the gate and the control films, thickness of the insulating films, etc., which determine the various characteristics such as input and output impedance, transimpedance, and time constant of the amplifier. For example, the time constant τ of the cascaded amplifier where the gate of one cryotron drives the control of another is given by

$$\tau = (L_c + L_g)/R$$

where L_c is the control impedance, L_g is the gate impedance, and R is the output gate resistance. Short time constants require L_c and L_g to be as small as possible. This can be achieved by the use of a superconducting shield plane separated from the control film by an insulating layer not thicker than that required to avoid electrical shorts and by making the control and the gate as wide as possible. It should be noted, however, that increasing the width of the gate, W_g, decreases not only the inductance but also the resistance of the gate, so that in actual practical circuits where $L_g \ll L_c$, increasing W_g increases the time constant. A small time constant therefore requires a wide enough control film and a narrow enough gate film, in contrast to just the reverse for obtaining high static gain of the switch. A compromise has to be made between the two requirements to obtain respectable values of gain and time constant. It has been found by Newhouse and Edwards[23] that the width of the control film must not be less than 100–200 times the separation distance between the control film and the ground plane to ensure maximum transimpedance and hence maximum dynamic gain per stage, defined as

$$G = R_t/R_g$$

where $R_t = (\partial V_g/\partial I_c)_{I_g}$ is the transimpedance and $R_g = \partial V_g/\partial I_g$ is the differential gate resistance. An optimum value of W_g is chosen on the basis of the required value of L_g, which does not play a dominant role in determining the time constant although it determines the frequency response of the amplifier. Generally, the frequency response of cryotron amplifiers is limited to less than 1 MHz due to Joule heating in the gate and poor heat transfer to the low-temperature bath. The width of the gate therefore need not be greater than that required to give response up to these frequencies. Once W_g is thus fixed, the thickness of the gate film can be adjusted to give the optimum value of the gate resistance which deter- mines both the gain and the response time. The thickness of the gate film must exceed a certain value to avoid broadening of the transition but should be less than $\sqrt{5}\,\lambda$ in any case to avoid hysteresis effects.[24]

Based on this analysis, fabrication of an eight-stage amplifier with a maximum current gain of 22,000, an upper cutoff at 45 kHz, and a minimum detectable input current (r.m.s.) of 6.8×10^{-8} Å has been reported.[25]

A simpler and more useful cryotron amplifier is the single-stage multi- crossover design developed by Newhouse *et al.*[26] It consists of a 200 μm wide, 3000 Å thick Sn or In gate film and a 200 μm wide, 7500–10,000 Å thick Pb control film, arranged in an array of 512 crossovers (as shown in Fig. 6.1f) deposited on a superconducting ground plane of Pb film evapor- ated on glass, single-crystal quartz or anodized Nb. Insulating films of SiO (4000 Å thick) are used between all metal films. The low-frequency noise produced due to local thermal fluctuations in this type of amplifier is much smaller compared to the multistage CFC amplifier. The unique features of cryotron amplifiers, such as an unusually low input-noise level, zero input resistance, and low input inductance, makes them very attractive for special applications.

6.4.3. Computer Memory Devices

A cryotron may be used as a storage device in two ways: by causing it to steer current from one superconducting path to another or by using it to cause persistent currents to flow in a superconducting loop. Thus binary information may be represented by the "presence" and "absence" of current or by "clockwise" and "anticlockwise" directions of current in a superconducting loop. Superconducting memory devices utilize the property that a current induced in a superconducting ring will persist indefinitely. The simplest type of a cryotron-based storage cell is the persistent current cryotron cell[27] shown in Fig. 6.2a. It consists of a superconducting loop with a low-inductance branch containing the gate of an input cryotron and a high-inductance branch containing the control of an output cryotron. Initially, when the entire loop is superconducting, a supply current I_0 will divide between the two parallel branches inversely as the ratio of their inductances, with the result that most of the input current I_0 flows in the low-inductance arm. On activating the control of

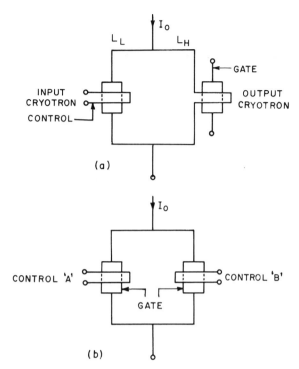

Figure 6.2. Two types of cryotron storage cells: (a) persistent-current cryotron-based cell; (b) cryotron flip-flop.

the input cryotron by applying a write current I_w to it, its gate goes resistive and the current is diverted to the high-inductance branch. When the input gate goes superconductive again upon deactivation of the control, switching off the current I_0 results in a clockwise persistent current of magnitude $I_0 L_H/(L_H + L_L)$, (where L_H is the inductance of the high-inductance branch and L_L, that of the low inductance branch) around the loop. The persistent current is destroyed by activating the input control in the absence of supply current I_0. The output, or read cryotron, gate is resistive or superconductive depending on whether there is or is not a current in its control which is a part of the storage loop. A logical "1" or "0" is thus defined by the presence or absence of a current in the superconducting loop.

Another storage cell where binary information is defined by the direction of the persistent current is the cryotron flip-flop shown in Fig. 6.2b. In this circuit gates of both the cryotrons are in the storage loop. Initially the supply current I_0 is made to flow in the arm A by temporarily opening gate B by passing a current through its control. Activation of gate A when the entire loop is superconducting diverts the current to arm B, and it remains there even if the current pulse on gate A is removed. The system therefore acts like a flip-flop. Storage is introduced in the cell by removing the supply current when a clockwise persistent current of magnitude $I_0 L_B/(L_A + L_B)$, if the current were originally in arm B, or an anticlockwise current of magnitude $I_0 L_A/(L_A + L_B)$, if it were in arm A (L_A, L_B are the inductances of arms A and B, respectively), is left circulating in the loop. The logical state of the element may be determined, for example, by using gate A current I_{gA} for driving the control of a sense cryotron, the gate of which develops a voltage if I_{gA} is finite, but not if it is zero.

The earlier noncryotron type of storage cells are the persistor,[28] persistaron,[29] and the Crowe[30] cell. In all these cells, switching is accomplished in a part of the storage loop directly by passing current through it, rather than by an externally controlled magnetic field. The persistor and the persistaron (Fig. 6.3) are somewhat similar in operation and are comprised of a superconducting loop consisting of a permanently superconducting inductance connected in parallel with a superconducting film of negligible inductance, where most of the supply current flows initially. The low-inductance branch goes resistive beyond a critical value of current passing through it, steering the current to the high-inductance branch in a time L/R, where L and R are the total loop inductance and resistance, respectively. An output signal appears across the low-inductance branch when it goes resistive. The time for which it remains in the resistive state after removal of the write current I_0 depends on the thermal time constant, which is a critical factor in the operation of these devices. Thus, although the basic principle is the same, several modes of operation are possible with various combinations of the polarity, magnitude, and shape of the write and read pulses. For example, if the thermal time constant (i.e., the time required by the film to dissipate the Joule heat and become supercon-

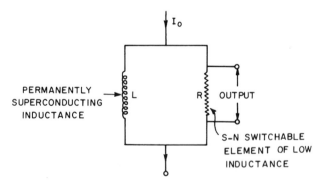

Figure 6.3. A noncryotron-type storage cell.

ducting again) is less than the electrical time constant L/R, as in the persistor, the low-inductance branch turns superconductive before the current is completely diverted to the high-inductance branch, so that, using a write pulse of duration slightly greater than L/R, a "1" is written on removal of the write pulse as a persistent anticlockwise current with magnitude equal to the value of the current in the high-inductance branch, at the time when R goes superconductive again. To write a "0" in the persistor, a negative pulse is used, which will store a clockwise persistent current even if a "1" had been stored there previously. The logic state of a persistor can be read by writing a "0." A voltage appears across R if a "1" was written previously, and no voltage appears if there was a "0." Since in either case the cell is left in the "0" state, the readout is destructive. By using inductive coupling of the drive coils, persistor elements can be used in a coincident current array for a random access memory (see Section 3.7). These devices have been successfully operated[31] at a 15-MHz repetition rate using 30-nsec pulses.

In cases where the thermal time constant is greater than L/R, the circulating current will decay to zero before the film becomes superconducting again, and therefore no current will be stored in the loop. In such a case it is necessary to apply a "primary" pulse at the beginning of the information-storage cycle.

The persistaron is similar to the persistor except that long rise time (compared to L/R) drive pulses are used. This makes heating effects almost negligible. Vail *et al.*[32] have operated a persistaron with 1-μsec pulses at a repetition rate of 16 KHz. The bridge cell[33] and the loop cell[34] shown in Fig. 6.4 are two illustrations showing how the above-described principles can be realized in thin film structures. In the loop cell, the hole shown in the ground plane serves to increase the inductance of the nongate side of the loop.

Another thin film storage cell is the Crowe cell,[30] shown in Fig. 6.5a. It consists of a thin hard superconducting film with a small hole ($\sim$ a few millimeters in diameter) bridged by a narrow soft superconductor crossbar which can be switched resistive when the current induced in it by magnetic coupling from two electrically insulated drive lines placed directly over and parallel to the crossbar exceeds the critical value. The current then returns to the edges of the hole. On reducing the drive current to zero, information is stored as a persistent current in the two D-shaped loops formed by each half of the hole and the crossbar. The output can be inductively sensed by a parallel sense line beneath the crossbar on the other side of the hole. Using a $600 \text{ Å} \times 0.1 \text{ mm} \times 2 \text{ mm}$ Pb crossbar deposited on various substrates, Broom and Simpson[35] reported time constants of their cells to be between 10 and 20 nsec. Rhoderick[36] obtained a 3 nsec time constant and an output voltage of $\sim$50 mV for his cells.

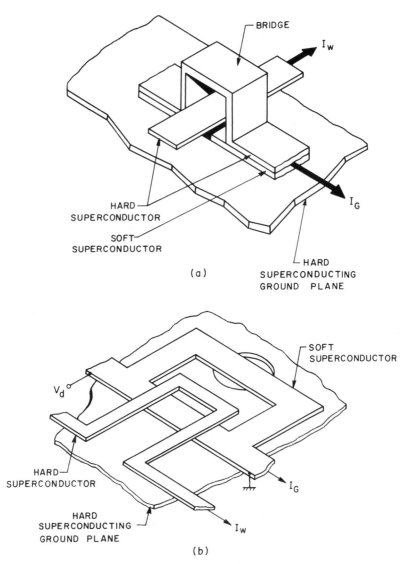

Figure 6.4. Bridge cell (after Ref. 33); (b) loop cell (after Ref. 34).

A memory cell of the continuous film memory (CFM) proposed by Buchold[37] is shown in Fig. 6.5b. It has no physical holes but consists of drive lines intersecting over a continuous film in such a way that the vector sum of the fields produced by drive currents "punches" holes of normal metal in the film at points where critical current is obtained. This produces a signal in the sense line located on the other side of the film.

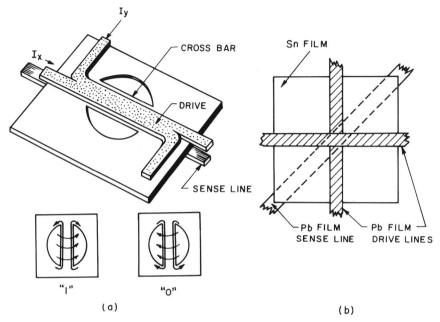

Figure 6.5. (a) The Crowe cell and its "1" and "0" stored modes (after Ref. 30); (b) continuous-film memory (CFM) cell (after Ref. 37).

So far we have described only individual storage cells. Various logic functions can be performed using suitable superposition of the drive currents. For many years, cryotron-based superconducting memory and logic devices enjoyed the reputation of being the promising future of high-speed computers, until they were superseded by a competing technology based on Josephson junctions, which are described in the next section.

6.5. *Superconductive Tunneling Devices*

Before going over to the discussion of the devices, let us briefly discuss the phenomenon of superconductive tunneling.

6.5.1. *Quasiparticle (Giaever) Tunneling*

The transport of electrons between two metal electrodes separated by a thin (≤ 50 Å) insulator barrier, by means of a quantum-mechanical tunneling process, is a well-known phenomena. The current–voltage (I–V) characteristic of such a M–I–M tunnel junction is ohmic at low bias voltages, as

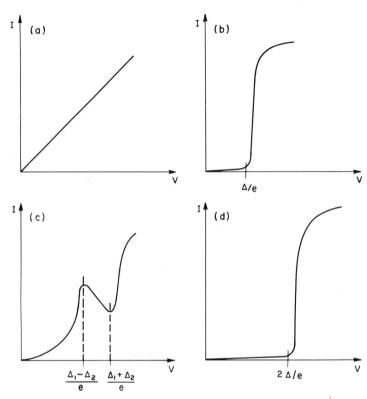

Figure 6.6. Current–voltage (I–V) characteristics of various tunnel junctions: (a) both metals normal; (b) one metal normal, one metal superconducting; (c) both metals superconducting; (d) two identical superconducting metals.

shown in Fig. 6.6a, and becomes nonlinear at high bias. The resistance of such a junction is nearly independent of temperature and increases exponentially with the thickness of the insulator layer. During his studies of tunneling in M–I–M junctions, Giaever[38] discovered that if one of the metals becomes superconducting (M–I–S), the I–V characteristics become highly nonlinear (Fig. 6.6b), with the current remaining nearly zero (exactly zero only at 0 K) for voltages up to $V = \Delta/e$ and rising sharply thereafter. It was later found[39,40] that if both the metals comprising the junction are superconducting (S_1–I–S_2) with energy gaps $2\Delta_1$ and $2\Delta_2$, a negative-resistance region is developed in the I–V curves for voltages between $|\Delta_1 - \Delta_2|/e < V < (\Delta_1 + \Delta_2)/e$, as shown in Fig. 6.6c. However, if both the superconductors are identical (S–I–S) with $\Delta_1 = \Delta_2 = \Delta$, the negative-resistance region disappears and the initial rise in current is observed only when $V = 2\Delta/e$, as shown in Fig. 6.6d. It is clear from the shape of the I–V

curves that a study of electron transport in tunnel junctions provides an accurate and remarkably simple method of determining the energy gap and density of states of a superconductor. The negative-resistance region in the I–V curve of a S_1–I–S_2 junction suggests its use for switching applications, as in amplifiers, oscillators, and computer elements. Miles *et al.*[41] used an Al–Al oxide–Pb junction at a temperature of ~ 1 K as an amplifier with a power gain of 23 dB at 50 MHz. Although in the beginning a large number of devices based on Giaever tunneling were proposed, interest in further research work on these devices was lost due to the complete overshadowing of these devices by the much more superior Josephson tunneling devices. For a comprehensive review of the Giaever devices, the interested reader is referred to Solymar's book[42] on superconductive tunneling.

6.5.2. Pair (Josephson) Tunneling

The remarkable discovery of Josephson[43] was the theoretical prediction that not only the normal single electrons (quasiparticles) can tunnel through the insulating barrier, but also the superelectrons (Cooper pairs), provided the barrier is thin enough (~ 10 Å). The insulating barrier acts as a weak link coupling the wave functions of the two superconductors on either side of the barrier. The criterion for a weakly coupled superconductor is that the Cooper pair density is substantially reduced locally with respect to the adjacent regions. A 100–1000 Å thick semiconductor layer or a 10,000 Å thick normal metal layer sandwiched between two superconductors (S–N–S) may also serve as a weak link. A low-pressure point contact between two superconductors is another example of a weak link. A weak link between the two regions of a superconducting film may be formed either by making a very narrow constriction (i.e., a region of width $<\lambda$) or by depositing an overlayer of normal metal between them. The former is called a *Dayem bridge*,[44] and the latter a *Notarys metal bridge*.[45] The term Josephson junction is used in its broadest sense to include all types of the above-mentioned weak links. The S–I–S type of junction consisting of a dielectric barrier is referred to as the *Josephson tunnel junction*. The various predictions of Josephson are described by the following three equations:

$$I = I_c \sin \theta \qquad (6.5)$$

$$d\theta/dt = (2e/\hbar)V(t) \qquad (6.6)$$

$$\text{grad } \theta = (2ed/\hbar)(\mathbf{H} \times \mathbf{n}) \qquad (6.7)$$

where I is the total junction current (for uniform current density); I_c is a constant (called the critical current) dependent on the energy gap (2Δ), the absolute temperature T, and the normal conductance G of the junction [$I_c = (\pi G/2e)\Delta(T)\tanh[\Delta(T)/2kT]$ for identical superconductors forming the junction]; θ is a gauge-invariant quantity given by $\theta = \theta_1 - \theta_2$, the difference between the phases of the quantum-mechanical wave functions in the superconductors S_1 and S_2 in the absence of any magnetic field, and by $\theta = \theta_1 - \theta_2 - (2e/\hbar)\int_1^2 \mathbf{A}\,\mathbf{dl}$, in the presence of a magnetic field $\mathbf{H} = \nabla \times \mathbf{A}$ with $\mathbf{A}$ as the magnetic vector potential. The integral is taken over a path joining two points lying deep inside (i.e., away from the link) S_1 and S_2.

In Eq. (6.6) V is the voltage across the junction, and in Eq. (6.7) d is the sum of the barrier thickness t and the penetration depths λ_1 and λ_2 in S_1 and S_2 ($d = t + \lambda_1 + \lambda_2$), and $\mathbf{n}$ is a unit vector normal to the plane of the junction.

All the Josephson effects are revealed by interpretation of the above three equations. For the case of a spatially and temporally constant nonzero value of θ, Eq. (6.5) states that in the absence of a magnetic field (that is, $\mathbf{A} = 0$) in the junction, a supercurrent I [it is a supercurrent because Eq. (6.6), for such a case, implies that $V = 0$] of maximum value $\pm I_c$ can flow through the junction without developing any voltage across it, and the insulating barrier behaves as if it were a superconductor! This effect is the so-called dc Josephson effect. If, on the other hand, the junction is subjected to a constant voltage V so that θ increases linearly with time [Eq. (6.6)], Eq. (6.5) implies that the supercurrent I starts oscillating with amplitude I_c and frequency $\nu_J = 2eV/h$ (where $2e/h = 483.6\ \text{MHz}/\mu\text{V}$). This effect, called the *ac Josephson effect*, is accompanied by emission of electromagnetic radiation of frequency ν_J in the microwave (gigahertz) region. This forms the basis of the use of Josephson junctions as sensitive and high-speed detectors and emitters of microwaves. For transport currents greater than I_c, the junction becomes resistive with a finite voltage appearing across it, and the current, for example, in the case of a tunnel junction, is then transported by Giaever tunneling. Since the normal current in this region is accompanied by an ac supercurrent, it is called the resistive–superconductive region.

The last of the three Josephson equations states that in the presence of a magnetic field (in-plane) in the junction, θ varies spatially in the junction plane in a direction perpendicular to the applied field, as a consequence of which the junction current I is nonuniform [Eq. (6.5)] along the junction. For such a case of spatially varying θ, Eq. (6.5) also indicates that the value of the maximum supercurrent that can flow through the junction decreases with increasing applied magnetic field. It can be

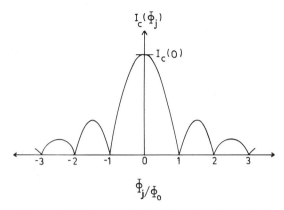

Figure 6.7. Dependence of critical current of a Josephson junction on the flux passing through it.

shown that

$$I_c(\Phi_j) = I_c(0) \frac{\sin(\pi\Phi_j/\Phi_0)}{\pi\Phi_j/\Phi_0} = I_c(0) \frac{\sin(\pi H/H_0)}{\pi H/H_0} \qquad (6.8)$$

where $I_c(0)$ is the value of the maximum permissible supercurrent in zero magnetic field; Φ_j is the total flux linking the junction ($\Phi_j = H \times$ area of the junction); Φ_0, as already mentioned, is a fundamental constant ($h/2e$) representing one quantum of flux; and H_0 is the field for which the junction contains one quantum of flux. Figure 6.7 shows the variation of I_c with Φ_j/Φ_0. This effect is called the *diffraction effect* for Josephson junctions because of the similarity of Eq. (6.8) to the expression for Fraunhofer diffraction from a single slit. It should be noted that Eq. (6.8) holds exactly only when the self-field of the junction generated by the current flowing through it is negligible in comparison to the externally applied magnetic field. If it is not so, the magnetic field dependence of the critical current may take a variety of symmetrical and asymmetrical shapes with respect to the I_c axis. It can be seen from Fig. 6.7 that minimal current values are obtained whenever the junction contains an integral number of flux units Φ_0. Since the area of a Josephson junction is typically $\sim 10^{-6}$ cm^2, a magnetic field ~ 0.1 G is required to produce one quantum of flux, showing that a single junction is not very sensitive to magnetic fields. The sensitivity can be greatly enhanced by connecting two identical junctions in parallel in a superconducting ring (as shown in Fig. 6.8a), which results in quantum interference effects between the two junctions. The tunnel junctions in

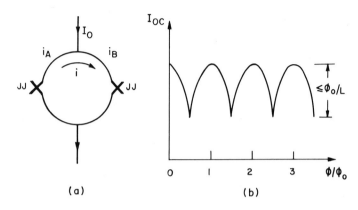

(a)   (b)

Figure 6.8. (a) A SQUID; (b) maximum supercurrent of a SQUID as a function of the flux passing through the ring.

particular need to be shunted with a resistance to eliminate hysteresis effects in the I–V curves (Fig. 6.9b).

From the principle of flux quantization in a ring, we know that if the loop were continuous (i.e., junctions are removed), the flux Φ inside the loop would be given by $\Phi = n\Phi_0$, where n is an integer and the corresponding phase shift in the wave function around the loop would be $2\pi n = 2\pi(\Phi/\Phi_0)$. In the presence of two Josephson junctions in the ring, this would be modified to

$$2\pi n = 2\pi(\Phi/\Phi_0) + \theta_A - \theta_B \qquad (6.9)$$

where θ_A and θ_B are the phase shifts across the two junctions. The total transport current I_0 through this device is then

$$I_0 = i_A + i_B = I_c(\sin\theta_A + \sin\theta_B) \qquad (6.10)$$

It should be noted that the flux Φ linking the area enclosed by the ring differs from the flux Φ_{App} applied to the ring, being related to it by $\Phi = \Phi_{App} + Li$, where L is the loop inductance and i is a circulating current in the ring, given by

$$i = (i_A - i_B)/2 = (I_c/2)(\sin\theta_A - \sin\theta_B)$$

(assuming that the inductance of each half-loop is $L/2$).

It can be shown from a solution of the above equations that the variation of the maximum supercurrent I_{0c} that the ring can carry as a

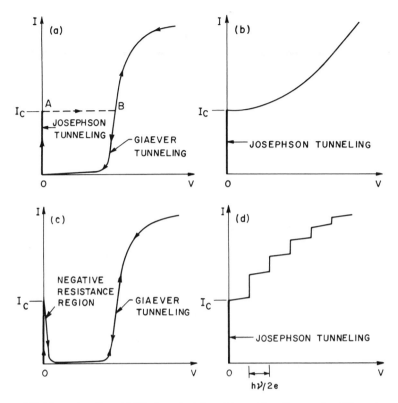

Figure 6.9. Current–voltage (I–V) characteristics of tunnel junctions under different conditions: (a) driven with constant-current source; (b) shunted with a resistance; (c) driven with constant-voltage source; (d) voltage-biased and irradiated with electromagnetic radiation of high frequency.

function of the flux in the ring is

$$I_{0c}(\Phi) = 2I_c(\Phi_j) \cos (\pi\Phi/\Phi_0)$$

$$= 2I_c \sin \frac{(\pi\Phi_j/\Phi_0)}{(\pi\Phi_j/\Phi_0)} \cos (\pi\Phi/\Phi_0) \qquad (6.11)$$

Since the area of junctions is orders of magnitude smaller than the ring area, the variation of I_{0c} is essentially given by $|\cos (\pi\Phi/\Phi_0)|$, as shown in Fig. 6.8b. The minimum value of I_{0c} which occurs at $\Phi = (n + \frac{1}{2})\Phi_0$ depends on a parameter $\beta_L = 2\pi L I_c/\Phi_0$, whereas the maximum value of I_{0c} occurring at $\Phi = n\Phi_0$ is equal to $2I_c$. In practice, although the periodicity of I_{0c} is always Φ_0, the predicted [Eq. (6.11)] modulation of I_{0c} from $2I_c$ to zero is not observed. Because of the dependence of $I_{0c,min}$ on L, the modulation

$(I_{0c,max} - I_{0c,min})$ in actual cases is limited to a value of $\leq\Phi_0/L$, as indicated in the figure. Since the area of a ring is much greater than that of a single junction, resolution of field changes $\sim 10^{-9}$ G is possible. This type of device employing quantum interference effects is called a SQUID (*s*uperconducting *q*uantum *i*nterference *d*evice). Different types of SQUIDs and their applications are discussed in detail in the next two sections.

Let us now look at the time-averaged current–voltage (dc I–V) characteristics of Josephson junctions, which may have different shapes depending on the source used to drive the junction and the presence of high-frequency electromagnetic noise. We have discussed earlier that, for finite voltages, current transport takes place by two mechanisms, one being the oscillating supercurrent, $I = I_c \sin \theta(t)$ and the other being the normal current given by GV, where G is the normal conductance, which may be ohmic (i.e., independent of V) as in the case of point-contact, S–N–S, and metal bridge types of Josephson junction, or a function of V as in the case of a tunnel junction. In tunnel junctions there is yet another current-carrying mechanism in addition to the above two. This is the displacement current through the insulator barrier due to the high junction capacitance. A real Josephson tunnel junction can therefore be visualized as an ideal Josephson junction shunted by a capacitance C and by its own normal conductance $G(V)$, so that the total junction current is given by

$$I = I_c \sin \theta + VG(V) + C(dV/dt) \tag{6.12}$$

It is clear from this equation that, for zero voltage, a dc supercurrent of maximum value I_c flows through the junction as in the ideal case. The dc I–V characteristic of a tunnel junction driven with a constant-current (i.e., high ac impedance) source is shown in Fig. 6.9a. It shows that for currents greater than the critical current, the junction switches from the zero-voltage superconducting state (point A) to the gap voltage (point B) and the current is transported by Giaever tunneling. On decreasing the current, the Giaever tunneling curve is followed introducing a hysteresis behavior in the I–V curve which gives it a bistable nature. Josephson junctions thus form ideal elements for computer and other switching applications. The time of switching, on the basis of Josephson frequencies, is expected to be in the sub-picosecond region, but, in practice, is limited to about 10–50 psec due to the large junction capacitance. The ac supercurrent is shunted through the large junction capacitance and does not contribute to the dc I–V characteristic except at very low voltages where the Josephson frequency is low. In point-contact, S–N–S, and metal bridge types of Josephson junctions, where the inherent junction capacitance is negligible, under current-biased conditions the ac supercurrent, being nonsinusoidal, does not time-average to zero, with the result that the I–V curve rises continuously and monotonically, as shown in Fig. 6.9b. Elimination of hysteresis in the $I-V$ curve of the tunnel junction, as required for SQUID applications, can be achieved by shunting the junction with a resistance such that the damping parameter, $\beta_c = 2eI_cC/\hbar \ [G(0)]^2$, is less than 1. The $I-V$ curve of such a shunted tunnel junction is similar to that of a point-contact junction, shown in Fig. 6.9b.

When a Josephson tunnel junction is driven by a constant-voltage (i.e., low ac impedance) source, the $I-V$ characteristic exhibits a negative-resistance region for currents less than the critical current I_c, as shown in Fig. 6.9c. The negative-resistance characteristic can again be utilized for switching applications as in signal switches, amplifiers, and oscillators. Stewart[46] and McCumber[47] have done a detailed analysis of the shape of various $I-V$ curves on the basis of equivalent circuits.

Another very important property of a Josephson junction is that, when it is voltage biased and irradiated with electromagnetic radiation of high frequency (ν), as, for example, microwaves, the beating of the two frequencies ν and ν_J produces a zero-frequency current component whenever the supercurrent frequency ν_J is an integral multiple of ν. As a consequence, a series of constant-voltage current steps is introduced in the dc $I-V$ characteristic (Fig. 6.9d) at voltages given by

$$V_n = nh\nu/2e$$

where n is an integer.

An accurate measurement of the microwave frequency and the voltages at which the steps appear provides by far the most accurate method of determining the fundamental constant h/e (which has now become known[48] to an accuracy of 0.12 ppm) as well as a standard for voltage in terms of a known frequency. The standard U.S. legal volt is now based on the Josephson effect. Since the measurement of e/h using this method is independent of any assumptions of quantum electrodynamics, it removes the well-known discrepancy between the theoretical and experimental values of the hyperfine splitting in the ground state of atomic hydrogen. This is a very important fundamental contribution of thin film phenomena to basic solid-state physics.

6.5.3. *SQUIDs*

It was shown in the previous section that a SQUID is extremely sensitive to magnetic fields. This property of SQUIDs can be exploited for their use[49] as incredibly sensitive magnetometers, gradiometers, fluxmeters, voltmeters, ammeters, amplifiers, etc. Obviously, it is not possible here to go into the details of the complicated electronic circuitry of these devices, although a brief mention of the principle of their operation is made in the next section. In the present section we describe how the various geometries of SQUIDs can be realized in thin films. However, before doing that we would like to mention another somewhat similar device called (incorrectly though), rf SQUID. (For distinction, the previously described double-junction structure is referred to as dc SQUID.) It consists of a low-inductance ($\sim 10^{-9}$ H) superconducting loop, interrupted by a single Josephson junction and inductively coupled to a high-quality tank circuit (as that in the input of an rf amplifier) which is driven by a weakly coupled rf current source. The rf voltage appearing across the tank circuit varies periodically (the period being Φ_0) with the flux applied to the superconducting loop, making the function useful for devices similar to those employing dc SQUIDs. An additional advantage of rf SQUIDs over the dc SQUIDs is their ease of fabrication.

We now describe the actual form of thin film SQUIDs for which two geometries, cylindrical and planar, are possible. One common geometry of commercially available thin film rf SQUIDs is shown in Fig. 6.10a. It consists of a 300 Å thick superconducting film evaporated on a cylindrical quartz rod (7 mm with 1.5 mm diameter) with a narrow Dayem bridge (0.5 μm wide and 5 μm long) scratched or etched in its circumference[50] as shown in the figure. Using Notarys normal metal overlay bridges, Notarys and Mercereau[51] fabricated rf SQUID magnetometers and other instruments using microelectronic techniques, with all the components evaporated on the same chip.

Planar rf SQUIDs with a microbridge and two holes were fabricated by Fujita *et al.*[52] by sputter etching of NbN thin films deposited epitaxially on MgO substrates. When operated at 15 MHz, a flux resolution $\sim 10^{-3} \Phi_0$ Hz$^{-1/2}$ (with inherent flux noise one order smaller), corresponding to a field sensitivity $\sim 7 \times 10^{-9}$ G, was obtained.

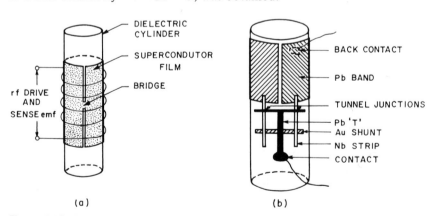

Figure 6.10. Common geometrics of (a) rf SQUID (after Ref. 50) and (b) dc (cylindrical) SQUID (after Ref. 53).

The most popular form of dc SQUID using cylindrical geometry and fabricated by Clarke *et al.*[53] uses Nb–NbO$_x$–Pb tunnel junctions. To fabricate this type of SQUID, shown in Fig. 6.10b, first a band of Pb (alloyed with 5% In) approximately 10 mm long and 3300 Å thick is evaporated onto a quartz cylinder (18 mm long with 3 mm diameter), and a slit parallel to the cylindrical axis is formed in it. A 250 μm wide and 1200 Å thick strip of gold (with a 500 Å thick underlayer of chromium to improve adhesion) is then evaporated to serve as a shunt for the junctions and eliminate hysteresis in the *I–V* curves. Next two vertical strips of Nb, 150 μm wide and 1400 Å thick and separated by 1 mm, are deposited with their one end touching the lead band. To form the junctions, a "T" of Pb with a 75 μm wide junction-forming arm is deposited after slightly oxidizing the Nb strips. A flux resolution of $3 \times 10^{-5} \Phi_0 \, \text{Hz}^{-1/2}$ was achieved with this SQUID by optimization of the electronic circuitry and careful shielding. Because of the presence of Pb films, the device needs a waterproof protective coating on it. A Nb–NbO$_x$–Nb version of this device is, however, extremely stable[54] to environmental effects and needs no protective measures.

6.5.4. Applications of SQUIDs

The use of SQUIDs as magnetometers has already been described in the above section. The magnetic field sensitivity of a SQUID can be increased by using a superconducting flux transformer, a schematic diagram of which is shown in Fig. 6.11a. It consists of two superconducting loops A and B (with area A_1 of loop A $\gg$ area A_2 of loop B) connected in series by superconducting wires. A magnetic field H_1 applied to the pickup loop A generates a supercurrent in the transformer which produces a magnetic flux Φ in a SQUID which is coupled to loop B through a mutual inductance *M*. For this case

$$\Phi = MA_1H_1/(L_1 + L_2)$$

where L_1 and L_2 are the self-inductance values of loop A and loop B, respectively. Therefore the magnetic field sensed by the SQUID, H_2, is

$$H_2 = MA_1H_1/A_2(L_1 + L_2)$$

Magnetic field sensitivity is enhanced since $A_1 \gg A_2$.

A SQUID can also be made to measure flux gradients by using the circuit shown in Fig. 6.11b. It uses a flux transformer with two equal opposed primary coils so that a uniform applied field does not produce a supercurrent in it. But if a field gradient is present, a supercurrent proportional to δH

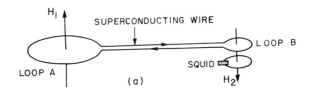

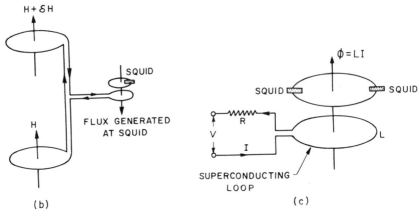

Figure 6.11. Schematic diagrams of a flux transformer (a), flux gradiometer (b), and voltmeter (c).

is induced in it which generates a flux proportional to δH at the SQUID through a coupling coil. The device thus measures difference in the fields and therefore acts as a gradiometer. A thin film dc SQUID gradiometer with a sensitivity of about $2 \times 10^{-10}\,\mathrm{G\,cm^{-1}\,Hz^{-1/2}}$, was fabricated by Ketchen et al.[55] by evaporating pickup loops of Pb and tunnel junctions of Nb–NbO$_x$–Pb on a single planar substrate of quartz.

The sensitivity of SQUIDs to magnetic fields also implies that they can be used to measure currents and voltages by measuring the magnetic fields produced by them. Although the sensitivity of SQUIDs as ammeters ($\sim 10^{-10}\,\mathrm{A\,Hz^{-1/2}}$) does not approach that of other measuring devices, these have the advantage of having truly zero dc input resistance. The schematic circuit of a SQUID voltmeter is shown in Fig. 6.11c. The current ($I = V/R$) that is induced in the superconducting loop generates a magnetic flux (proportional to the voltage) in the SQUID. Although, in principle, the voltage resolution of a SQUID voltmeter is $\sim 10^{-17}$ V, in practice it is limited to $\sim 10^{-15}$ V by Nyquist noise.

We will now mention some of the areas of applications requiring the use of SQUIDs to measure magnetic fields. The most talked about application of SQUIDs is in biomagnetism, a subject of great medical interest, for measuring the tiny magnetic fields ($\sim 10^{-7}$ to 10^{-11} G) produced by the biological activity of various organs. The wide spectrum of signal measurements that have been made possible with SQUID magnetometers and gradiometers include magnetocardiogram (MCG) from heart activity, fetal MCG from heart activity in the fetus, magnetomyogram (MMG) from muscular activity, magnetoencephalogram (MEG) from activity of neurons in the brain (alpha rhythm), magnetooculogram (MOG) from movement of the eyes, and visually evoked fields (VEF) from visually stimulated brain activity. For a full review on the role of SQUIDs in this field, we refer the reader to Refs. 56–59.

SQUID susceptometers have been used for investigation of electronic[60] and nuclear[61] magnetism in solids at low temperatures. SQUIDs have also been used in millikelvin thermometry,[62] in gravity-wave-detection experiments,[63] in detection of asbestos in human lungs, and in many other applications.[49] Another application is the tracking of magnetic objects. An array of three SQUID magnetometers and five gradiometers along with sophisticated processing techniques was used by Wynn *et al.*[64] to track a moving magnetic monopole. SQUIDs have also proved very promising for geophysical explorations, such as for measuring fluctuations in the earth's magnetic field and local geomagnetic fields for mineral surveys.[65] SQUID gradiometers can be used for detecting slow changes in the magnetization of rocks which occur due to piezomagnetism and are the precursors of some seismic disturbance.[66] This may therefore ultimately be used for the prediction of earthquakes. A general overview of SQUIDs and their applications can be found in greater detail in several review articles[67–71] by various authors.

6.5.5. *Superconducting Electronics*

The new technology of superconducting electronics, employing Josephson tunnel junctions and SQUIDs as basic components, although still in its infancy, is becoming a challenge to the well-established Si technology in ultrahigh-performance electronic devices and systems. It combines extremely fast switching and short logic delays with incredibly small power dissipation, thereby allowing miniaturization and high circuit densities which are limited in Si technology due to problems associated with the removal of increased amounts of dissipated heat as a result of miniaturization. As compared to Si logic circuits which are limited to about 1 nsec logic delays with 10 mW of power dissipation per gate, logic circuits employing Josephson junctions typically have about 100 psec logic delays with about 10 μW power dissipation. The order of magnitude ($\sim 10^{-18}$ J compared to a room-temperature thermodynamic fluctuation energy of $\sim 10^{-21}$ J) of the energy associated with Josephson junctions (the coupling energy $E_j = \Phi_0 I_c \cos \theta / 2\pi$) indicates that superconducting electronics is intrinsically low-energy-responsive electronics. The nonlinearity effects in the dc I–V characteristics of Josephson junctions, which are a key to the successful accomplishment of various basic functions in electronics, occur at very low energies ($\leqslant 5$ meV or 10^{-22} J). For these nonlinearities to be effective, the thermal fluctuation energy should be much smaller than this value, otherwise the nonlinearities tend to be averaged out. This requires the junctions to be operated at temperatures much below T_c. Let us now see how the basic electronic functions of signal detection, amplification, logic, and memory can be achieved using Josephson junctions.

A Josephson tunnel junction biased on the quasiparticle nonlinearity of the I–V characteristic (Fig. 6.12a) serves detection purposes in the same way as a diode does by rectification. A current responsivity of 3500 mA of output power per watt of input power at 36 GHz and a noise-equivalent power (NEP) of 2.6×10^{-16} W Hz$^{-1/2}$ have been reported by Richards *et al.*[72] This responsivity corresponds to an output of 0.52 electrons per incident photon, a value within a factor of 2 of the quantum limit of one electron per photon. In another experiment, a thin film SQUID of cylindrical geometry was used by Nisenoff[73] to detect electromagnetic radiation of frequency from 1 KHz to 10 GHz, with a predicted sensitivity of $\sim 10^{-17}$ W Hz$^{-1/2}$ for millimeter and submillimeter radiation. It has been shown by Shen *et al.*[74] that a superconducting heterodyne mixer operating at 36 GHz can very nearly detect a single photon.

Amplification also can be carried out using SQUIDs. A SQUID amplifier is nothing but a SQUID magnetometer tightly coupled to an input coil, as shown in Fig. 6.12b. Amplification is achieved by magnetic modulation of the critical current of the SQUID by the input signal. In this case also the energy sensitivities per unit bandwidth are slowly approaching the theoretically predicted[75] value of $h/2$, which is very close to the maximum energy sensitivity (h) per hertz that any quantum-

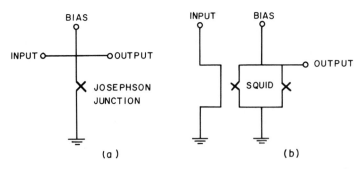

Figure 6.12. Schematic diagrams of a SQUID detector (a) and a SQUID amplifier (b).

mechanical system can have in accordance with Heisenberg's uncertainty principle.

Before moving on to logic circuits, it may be mentioned once again that a Josephson tunnel junction can be switched from a zero-voltage to a finite-voltage state by applying to it a current greater than the critical value. The time of switching depends on the device capacitance and the bias current and can be made extremely short. Binary information "0" and "1" can thus be represented as zero-voltage and finite-voltage states.

An actual Josephson junction switch[76] takes a form similar to an in-line cryotron except that the gate is a tunnel junction, as shown in Fig. 6.13a. The magnetic field of the current I_C in the control (which lies above the gate but is insulated from it) modulates the critical current I_{Gc} of the gate [which is current biased at somewhat below its critical current (I_C) at zero control current] and switches it to a finite-voltage state when the threshold current I_{Gc} is reduced below the gate bias current. Figure 6.13b shows the threshold characteristic of this device. Here, $n = 0$

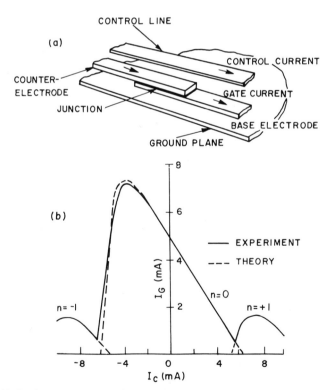

Figure 6.13. In-line Josephson junction switch (a) and its threshold characteristics (b) (after Ref. 76). I_C: control current; I_G: critical gate current.

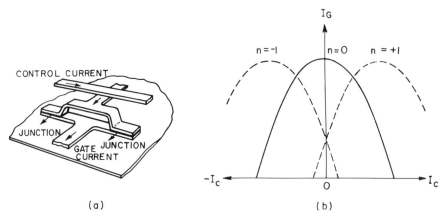

Figure 6.14. Two-junction switching device (a) and its threshold characteristics (b) (after Ref. 93). I_C: control current; I_G: critical gate current.

denotes the threshold curve valid when no quantum is trapped in the junction, and $n = \pm 1$ denotes curves which are valid when one flux quantum is trapped, with the $+$ and $-$ signs referring to the flux direction. A corresponding two-junction switching device[77] along with its threshold characteristics is shown in Fig. 6.14. Three modes of operation, namely, the self-resetting mode, the nonlatching mode, and the latching mode, are possible by the use of a proper load line. In the self-resetting mode, the device by itself reverts to zero once its current is transferred to the load, irrespective of the presence or absence of the control current. In the nonlatching mode, the device returns to the $V = 0$ state only when the control current is removed. In the latching mode, the device remains in the finite-voltage state even if the control current is removed and returns to zero only when the junction (i.e., gate) current is also removed momentarily.

One way of performing logic operations is[78] by using these switches in superconducting loop structures as with cryotrons. Although loop structures (Fig. 6.15a) are very useful for storage of binary information as persistent circulating currents, they do not provide very fast logic operation. Another configuration[79,80] which is particularly suitable for fast logic circuits is shown in Fig. 6.15b. A load resistor R_L shunts a current-biased Josephson junction which is controlled by control lines. If R_L is much smaller than the resistance of the junction in the finite-voltage state, most of the bias current will be diverted to R_L, which then can act as a control for the output sense junction. Using multiple control lines, various logic operations can be carried out by suitable superposition of the control currents. One of the fastest logic circuits reported by Gheewala[81] is a two-input OR gate with a switching delay of 13 psec and power dissipation

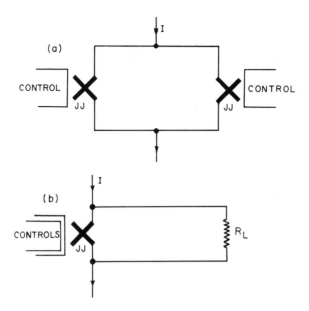

Figure 6.15. Two configurations for performing storage and logic using Josephson junctions: (a) loop structure for memory and slow logic (after Ref. 78); (b) configuration for fast logic operation (after Ref. 79).

per gate of $3\,\mu W$, fabricated using $2.5\,\mu m$ linewidth lithography. Such circuits provide[82] a basis for a computer with 2-nsec logic cycle time. A complex logic circuit consisting of a total of 51 interconnected multicontrol circuits and implementing a 4 bit × 4 bit add-and-shift multiplier, with a calculated multiplication time of 12 nsec (measured time 27 nsec, due to limitation of instrumentation), has been reported by Herell.[83] A shift register with shift rates of 160 Mbit/sec has also been reported.[84]

In addition to storage of information by flux trapping in superconduct-ing loops, it is also possible[77,85,86] to store information as the presence or absence of a single quantum of flux Φ_0 in the gate of the switching device itself (instead of the loop). This makes use of the large overlapping of the threshold curves for the $n = 0$ and $n = \pm 1$ states (Figs. 6.13b and 6.14b), particularly in the double-junction structure. The curves show that no flux quantum can be trapped in the gate at zero control current. However, if the bias conditions are made such that the point (I_G, I_C) lies within the, say, $n = +1$ threshold but outside the $n = 0$ threshold curves, a flux quan-tum is trapped. If the device is biased in the overlap region of the threshold curves for $n = 0$ and $n = +1$, the flux quantum would be retained in the device if the point (I_G, I_C) were originally in the $n = +1$ region, and there would be no flux quantum if it were originally in the $n = 0$ region. Thus by appropriate superposition of the gate and the control currents, the device can be made to switch to a finite-voltage state if it has a trapped flux quantum, and not to switch if there is no flux quantum. Switching times ~20 psec have been observed in these devices.

Zappe[87] reported loop-type memory cells of 50 μm linewidths with 600 psec write time and more than 5×10^8 NDRO operations without loss of flux and information. Loop memory cells with 2 μm linewidths with an upper limit of less than 100 psec write time have also been reported.[88] A fully decoded 64-bit NDRO random access memory array with read cycles ~5 nsec has been fabricated and tested by Henkels and Zappe.[89]

Due to limited space, we have attempted to describe only some of the highlights of superconducting electronics. Detailed information on this exciting subject of superconducting electronics may be found in some of the excellent reviews and papers on the subject.[90–93]

6.6. Miscellaneous Applications

In this section, some applications which have not been covered in the previous sections are briefly mentioned. The use of superconducting thin films for bolometric radiation detectors is mentioned in the next chapter along with other thermal applications. Use of Josephson junctions for thermometry was first suggested by Kamper[94] and later exploited by Kamper and Zimmerman[95] using Nb–NbO$_x$–Nb tunnel junctions and Nb–Nb point contacts. Here, a Josephson junction is voltage biased at $V \ll \Delta/e$ by connecting across it a small resistance $R \ll R_n$ (R_n is the normal resistance of the junction) and current biasing the combination. The ac Josephson frequency is modulated by the Josephson noise voltage generated by this resistance R at temperature T and as a result has a linewidth given by

$$\Delta\nu = 4\pi k TR/\Phi_0^2$$

The absolute temperature T can thus be measured by an accurate measurement of the linewidth of the radiation emitted by the junction if R is known accurately. Temperatures as low as 1 mK should be feasible. Using this technique other noise and fluctuation phenomena, such as fluctuations in the resistance of a superconductor near its critical temperature, can also be measured.

Some of the other applications of Josephson junctions are their use[96] in frequency mixers, radiation detectors, and parametric amplifiers, for the generation and detection of phonons,[97] and as particle detectors.[98] Particle detectors utilize transition edge-biased films and tunnel junctions similar to bolometers and sense the effective increase in temperature due to the loss of energy by the particle in the detector. Ar, Ar^+, He, and He^+ particles with energies between 150 and 800 eV can be detected by using a 1 μm wide and 500 Å thick Sn–In film held at 0.001 K below T_c. Similar applications have also been reported for molecular particles[99] and high-energy α-particles.[100] A tunnel junction such as Sn–SnO$_x$–Sn, voltage biased at $V < 2\Delta/e$ and $T \ll T_c$, serves as a MeV α-particle detector by measuring the quasiparticle current pulses produced by the particles.

Since all the currents and magnetic fields in superconductors are confined to a thin surface layer, superconducting thin films can be used to simulate bulk materials, in applications where cost and other considerations such as mechanical strength and thermal conductivity, etc., demand the use of some material other than the superconducting one. Therefore the required function of a superconductor can be performed simply by coating the base material with a thin ($\leqslant 1$ μm) superconducting film. For example, Pb-coated manganin wires (further covered with insulation) provide low-thermal-conductivity, resistanceless and easily solderable wires for interconnections in cryostats. Similarly, plastic ribbons coated with high T_c and high H_c materials such as Nb$_3$Sn retain the flexibility of the plastic ribbon even though the bulk alloy is brittle. These can then be wound easily to form high-magnetic-field superconducting solenoids. The use of superconducting-film-coated materials for magnetic screening is well known. Cavities and waveguides coated with superconducting films have very high values ($\sim 10^{10}$) of quality factor and find application in high-energy particle accelerators.

In conclusion, one can say that the unique, macroscopic quantum phenomenon of superconductivity has given rise to a wide variety of quantum-electronic devices. Although a number of applications and devices have not crossed the level of academic curiosity, the ones which have reached a sophisticated level of commercialization are already playing a very significant role in science and technology. The field of quantum engineering is indeed expected to have a bright future.

7

Thermal Devices

7.1. Introduction

It was not known until the discovery by William Herschel in 1800 that heat is a form of electromagnetic energy. Radiation in the infrared (IR) region with wavelength ranging from about 0.8 to 4000 μm constitutes that part of the electromagnetic spectrum which shows the greatest heating effect. Ever since this discovery, a large number of useful devices[1] have emerged. One very important class of these devices used for detection and imaging purposes is based on the fact that all bodies at temperatures above absolute zero emit radiation in the IR region which can be detected using suitable detectors.

Detectors in the IR region fall into one of the two categories (1) photon detectors and (2) thermal detectors. In photon detectors, the interaction of the individual photons above a certain threshold energy with the electrons in the material of the detector results directly in charge-carrier excitation or electron emission, without any intermediate process such as heating being involved. In thermal detectors, on the other hand, absorption of the incident radiation raises the temperature of the detector material, causing a change in the temperature-dependent properties of the material, for example, the resistivity of a metal or semiconductor (bolometric effect), the polarization of certain dielectric materials (pyroelectric effect), etc. In Chapter 3 we discussed in detail different types of photon detectors. In the present chapter, we will describe the principles of various types of thermal-detection and imaging devices and their applications.

The heating effect of radiation can also be used for direct conversion of electromagnetic energy into thermal energy. Efficient use of the large amount of freely available solar energy for heating and cooling purposes can be made using the principle of photothermal conversion. A detailed discussion of the photothermal applications of thin films constitutes the last part of this chapter.

7.2. Thermal Detectors

As already mentioned, thermal detectors operate by measuring the change in a temperature-dependent property of the material caused either by the absorption of IR radiation or by direct heating from a heat source. Since a large number of properties of thin films undergo rapid changes with temperature, different varieties of thermal detectors are possible depending on the property being exploited for detection. Thermal devices based on bulk materials suffer from two major limitations: (1) size, and (2) performance characteristics. The inherent ability of the thin film formation process to create unusual structures and compositions not found in the bulk and the possibility of having controlled IR absorption characteristics and low thermal capacity make thin film devices very sensitive and fast in resposne as compared to the bulk devices. Note that since a thermal detection process involves heating and cooling of the material, thermal detectors as a class are inherently slower in response as compared to photon detectors. Also, since thermal detectors respond to temperature changes caused by the integrated absorption of radiant energy, their response is wavelength independent over the spectral region where absorption is constant. Some of the common applications of thermal detectors include burglar and fire alarms, radio astronomy, tumor detection, etc.

Like any other detector, a thermal detector is characterized by its responsivity, detectivity, and noise-equivalent power, which are defined as follows:

1. Responsivity, R, is defined as the signal generated per unit incident power.
2. Detectivity, D, is defined as the signal-to-noise ratio per unit incident power and has the unit of W^{-1}. Since D is dependent on the area of the detector and the frequency bandwidth, thermal detectors in general are characterized by another quantity D^* (sometimes also called detectivity), defined as

$$D^* = D \times (AB)^{1/2} \, W^{-1} \, Hz^{1/2} \, cm$$

where A is the area of the detector and B is the frequency bandwidth over which the signal-to-noise ratio is measured. This definition of detectivity assumes a hemispherical field of view.

The noise-equivalent power (NEP) is related to D^* as

$$NEP = \sqrt{A}/D^* \, W \, Hz^{-1/2}$$

7.2.1. *Bolometers and Thermometers*

One of the most common thermal detectors is the one based on variation of electrical resistance with temperature. The device is called a *bolometer* if the change in temperature is brought about by absorption of IR radiation, and a *resistance thermometer* if the change is brought about by direct contact with a heat source. A resistance thermometer can be converted to a bolometer by depositing a suitable IR-absorbing film on it. Thin films of materials having high values of temperature coefficient of resistivity (TCR) are used as resistive elements of bolometers. During operation, a precisely controlled bias current i from a source with a regulating impedance is passed through the detector element of resistance r and TCR α, as shown in Fig. 7.1a. A voltage appears across the detector element when IR radiation falls on it. The voltage responsivity R of a bolometer is given by[2]

$$R = \eta i \alpha r (G^2 + \omega^2 H^2)^{-1/2} \tag{7.1}$$

where η is the fraction of incident power absorbed by the film, ω is the

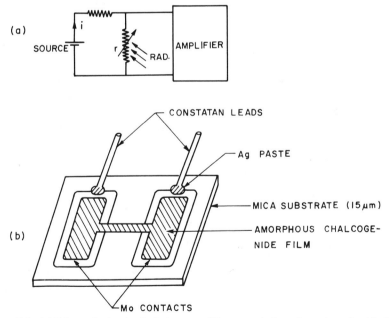

Figure 7.1. (a) Schematic circuit of a bolometer; (b) geometrical configuration of a thin film bolometer (after Ref. 7).

modulation frequency of the incident radiation, H is the thermal capacity of the detector, and G is the thermal impedance coupling the detector to its surroundings. Equation (7.1) shows that for high values of responsivity, high values of η, i, α, and r, and low values of G and H, are desired. It is clear from Eq. (7.1) that the input impedance of amplifier should be high as compared to r. Thus r cannot be increased independently. Also high values for r would mean increased Johnson noise, which arises in any resistive element due to random motion of the current carriers. The Johnson noise voltage for a given element of resistance r is directly proportional to the square root of r. The bias current i also cannot be increased beyond a certain limit because of intolerable noise due to Joule heating of the detector element. The remaining two parameters η and α can be maximized by choosing a suitable material. Low values of H can be achieved by making the detector as small and light as possible. This makes thin film bolometric elements much superior to bulk elements. The implication of having a small value of G is that the detector should, as far as possible, be thermally isolated from its surroundings. It should be noted here that this requirement of having low values of H and G distinguishes a bolometer from a thermometer.

Thin films of metals, semiconductors, and superconductors can be used as bolometric elements. For metal films, the TCR depends[3] on the film thickness, impurities, and decomposition conditions. Bolometers employing semiconductor films, also known as *thermistors*, are more sensitive compared to those employing metal films, because the TCR of semiconductor films is much larger. It depends on the band gap of the semiconductor, impurity states, and the dominant conduction mechanism. In the instrinsic region of conduction, for a semicondutor of electrical band gap E, the TCR at temperature T is given by $-E/kT^2$ (k is Boltzmann's constant). Thus, higher values of TCR are obtained with larger-band-gap materials and at low temperatures. The high resistivity of such a detector, however, makes it necessary to work with a material of intermediate band gap. Because of the high concentration of a variety of structural defects and also gaseous impurities invariably present in vapor-deposited films of semiconductors, conduction in such films is impurity and defect dominated. However, when depositied in amorphous form, films of most of the elemental as well as compound semiconductors have electrical conductivity orders of magnitude lower than the corresponding crystalline semiconductor due to very low free-carrier mobilities and behave[4-6] essentially like intrinsic semiconductors exhibiting no extrinsic conductivity regardless of the level of impurity content. This greatly simplifies the procedure for preparing a sample of desired electrical conductivity because it is the composition of the alloy that determines the electrical conductivity and not its impurtiy content, which plays a very dominant role in crystalline semiconductors. Also, their electrical conductivity is thermally activated with a single activation energy over a temperature range corresponding to a resistivity variation of a few orders of magnitude. All these factors make amorphous semiconductors more advantageous over the corresponding crystalline ones. One main disadvantage of these amorphous semiconductor bolometers is their high resistance and hence high Johnson noise.

Bolometers using chalcogenide glass elements of $10 \, \mu m$ thick sputtered amorphous films of $Tl_2Se \, As_2Te_3$ (on mica substrates $\sim 15 \, \mu m$ thick) overcoated with SiO have been fabricated by Bishop and Moore.[7] Using a 500 K blackbody source at 10 Hz chopping frequency, a $2.5 \times 10^{-3} \, cm^2$ device yielded a detectivity of approximately $2 \times 10^7 \, W^{-1} \, Hz^{1/2} \, cm$. The ac performance of the device is limited by its inherently long response time (~ 1 sec) because of the very low thermal conductivity of this glass. Figure 7.1b shows a schematic diagram of this device.

A similar device with better detectivity (5×10^6 to $5 \times 10^7 \, W^{-1} \, Hz^{1/2}$ cm, using a He–Ne laser as excitation source) and response time (1–10 msec) was obtained[8] using $25 \, \mu m$ thick sputtered amorphous $Ge_x H_{1-x}$ films on glass or sapphire substrates. Yoshihara[9] fabricated a bolometer using an amorphous Ge film detached from its substrate and claimed a detectivity of $\sim 10^8 \, W^{-1} \, Hz^{1/2} \, cm$ with a response time ~ 1 msec.

Semiconducting polymers such as PVC, PVAc, and PAN can also be used as bolometeric materials since they exhibit large changes in resistivity with temperature. Detectivities of $\sim 10^9 \, W^{-1} \, Hz^{1/2} \, cm$ have been claimed[10,11] for a 500 K blackbody source using a self-supporting PAN film backed by a black-gold layer. The performance of a bolometer can be improved by orders of magnitude by operating it at very low temperatures, firstly because the specific heat is much smaller and secondly because the resistivity changes are much larger compared to the room-temperature values.

Superconductive bolometers employ a thin film of a superconductor as the detector element. The film is biased into the intermediate state and is held at a temperature corresponding to the steepest part of the resistivity transition curve. Any radiation incident on the film which raises its temperature increases the resistance markedly. Because of the narrow superconducting transition, the operating temperature of this bolometer must be controlled to within $\sim 10^{-4}$ K. A recent suggestion[12] is to use a biasing magnetic field, thus varying the superconducting transition temperature to match the fluctuations in the bath temperature. Measurement of the resistance of the detector is done by sensing the penetration of the flux through a thin film rather than by passing a current. This helps to realize the full sensitivity of the device. Greater sensitivities are expected to result if these bolometers are used in conjunction with cryotron output amplifiers. Superconducting films of Sn on anodized aluminum substrate have been utilized by Gallinaro and Varona[13] to fabricate bolometers having a noise-equivalent power NEP $\sim 10^{-13}$ W Hz$^{-1/2}$ with response time $\sim$3 μsec. Superconducting bolometers may also find application in imaging devices and for the detection of microwaves and alpha particles. Thin metal films deposited on insulator substrates can also be used as high-frequency (up to 10^{10} Hz) thermometers[14] for studying the emittance of second sound in solids for other technical applications. Indium films about 2500 Å thick deposited on sapphire or quartz substrates exhibit[15] thermal relaxation times of about 10^{-9} sec in the temperature range 4–300 K. A high sensitivity to resistance changes has been achieved by operating the film very near to its superconducting transition temperature.

7.2.2. Thermocouples and Thermopiles

It was discovered in 1821 by Seebeck that in an open circuit consisting of two junctions of two dissimilar conductors (generally metals) different heating of one of the junctions generates a voltage which is a measure of the temperature difference between the two junctions. This is called the *thermoelectric* or *Seebeck* effect and forms the basis of this class of thermal detectors. Such a circuit between two dissimilar conductors is called a *thermocouple*. A thermopile is an array of a large number of thermocouples to provide a large output signal.

Thermopiles have been used[16] since 1833 to measure radiation energy. The basic element of a thermopile is a junction between different materials having a large Seebeck coefficient. Efficient performance of a thermopile requires (1) a large electrical conductivity to minimize losses due to Joule heating, (2) a small thermal conductivity to minimize conduction losses between the hot and the cold junctions, (3) a small thermal capacity for faster response, and (4) a high absorption coefficient for the incident radiation. We know that requirements (1) and (2) are not compatible in a material. The best compromise is obtained for some heavily doped semiconductors. For increasing the absorption of incident radiation, a thermopile can be coated at hot junctions with a radiation-absorbing film, whose spectral absorption characteristics determine the spectral characteristics of the thermopile. A blackbody coating yields a flat spectral response.

Thin films are ideally suited for thermocouple and thermopile applications. By using thin film and photolithographic techniques, a high density of packing of thermocouples in a variety of geometrical configuration can be conveniently fabricated. Figure 7.2 shows a 40-thermocouple pile of Ge : Cu–Constantan couples made in our laboratory. The common hot junction is coated with carbon black. The cold junctions are isolated and coated with a reflecting film of Al to protect them from the incident

Figure 7.2. A 40-element thin film thermopile made of Cu-doped Ge/Constantan thin film thermocouples fabricated in the authors' laboratory.

radiation. Temperature differences of less than 1 mK can be easily detected with this thermopile.

The sensitivity of a thermopile depends on the temperature difference ΔT per unit incident energy and the thermoelectric power (TEP), i.e., the Seebeck coefficient of the materials used for the thermocouples. The temperature difference ΔT can be maximized by minimizing the thermal capacity of the absorbing junction and the heat losses, and by suitable choice of the film thickness, the film, and the substrate materials. The TEP of a thin film is, in general, different from that of the corresponding bulk materials and depends on the film thickness. Also, the TEP is markedly affected by structural defects and disorder and by alloying impurities.[17-19] The relaxed solubility conditions in a vapor deposition process allow new materials with large TEP having a negligible temperature variation (see Fig. 7.3) to be prepared. For example, the TEP of highly disordered

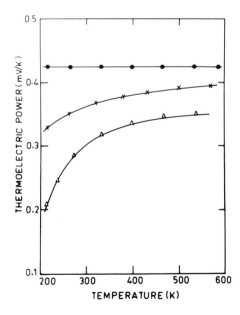

Figure 7.3. Temperature variation of TEP of some Ge-alloy films: (●) amorphous Ge–Al (2 at.%); (×) crystalline Ge–Fe (1 at.%); (△) crystalline Ge–Cu (5 at.%) (after Ref. 1).

vapor-quenched films of Cu–Ag is about 25 times larger than that of the corresponding crystalline films.[19] Anomalously large values (up to $5\,\text{mV K}^{-1}$) of TEP are obtained in amorphous films of Ge alloyed with metals such as Al, Cu, and Au.[17,18] Thermopiles of such materials, therefore, have a high sensitivity over a wide temperature range. Sensitivities of 5–$10\,\text{V W}^{-1}$ are easily attainable. It should be noted that whereas bulk wire thermopiles ordinarily used as, for example, in spectrophotometers can detect radiation down to $10^{-7}\,\text{W cm}^{-2}$ only, a thin film thermopile can detect radiation as low as $10^{-10}\,\text{W cm}^{-2}$. Also, whereas the response time of a fine wire thermocouple is a fraction of a second, that of a suitably designed thin film metal thermocouple can be[20] as low as $10^{-8}\,\text{sec}$. Multi-element thermopiles are slower, with response time of about a second for wire thermopiles and of several milliseconds for a thin film device.

Thin film thermopiles are ideally suited as high-speed thermometric and radiation measurement devices. Such devices are now commercially available and have arrangements (using two identical thermopiles) to compensate for changes in the ambient temperature. An array of microminiaturized thin film thermocouples can be used to detect and locate precisely an IR source such as hot gases emitted from a moving airplane or a missile. Such devices find major applications in defence. We have also utilized thin film thermocouples in a thin film differential thermal analyzer (DTA) to study structural changes and transformation in microgram specimens. Structural changes in a material are accompanied by absorption or evolution of a quantity of heat characteristic of the change in the lattice order. An effective method of studying these changes is to deposit the film under study directly onto the substrate on which the thin film thermocouple is deposited.

7.2.3. Pyroelectric Detectors

The pyroelectric effect is exhibited by certain low-crystalline symmetry materials which, due to lack of a center of symmetry, possess a spontaneous internal electric dipole moment. At a given temperature, the internal charge distribution is neutralized by an extrinsic charge distribution near the surface of the material, so that no voltage can be detected between the two electrodes attached to the opposite faces of the sample. If, however, the temperature is rapidly changed, a measurable change in the surface charge is produced due to changes in the internal dipole moment. This occurs because for good pyroelectric materials (which are good insulators) the extrinsic charge distribution is comparatively more stable. This effect can therefore be exploited for fabrication of a sensitive detector of modulated radiation operating at room temperature. A pyroelectric detector is a capacitor with two electrodes attached to the opposite faces of the temperature-sensitive material. A change in temperature or incident thermal radiation is detected by measuring the charge on the condenser. An ideal material for such a detector should have a large pyroelectric coefficient, a high absorption coefficient for the incident radiation, a high resistivity, a small dielectric constant, a low thermal capacity, and a low thermal conductance. Under ideal conditions, temperature changes of the order of 10^{-6} K can be measured with pyroelectric detectors. Pyroelectric films of $BaTiO_3$, $LiNbO_3$, and $LiTaO_3$ have been used[21] as sensitive detector elements with a response time of 1 μsec.

7.2.4. Absorption-Edge Thermal Detectors

The shift in the sharp absorption edge of a suitable semiconductor with temperature can be used as a thermal detection process. The shift in the absorption edge is observed by the change in the transmission of light of a suitable wavelength, as shown in Fig. 7.4.

Thermally induced structural changes at the molecular level take place in certain thermoplastics and liquid crystals. The changes modify the optical scattering and absorption processes and thus the color of thin films of these materials. A wide range of temperatures can be detected sensitively by employing thin films of appropriate liquid crystal and thermoplastic materials.

In addition to the above-described detectors, some other thermal effects can also be utilized for thin film thermal detectors. For example: (1) by keeping a ferromagnetic-based material near the curie temperature, a small change in temperature due to absorption of incident radiation results[22] in a sharp change in magnetization which can be measured magnetically or magneto-optically as described in Chapter 5; (2) the strong temperature dependence of the coercivity of materials such as permalloy, CoP, PdCo, etc. may be exploited for thermometric application; (3) the

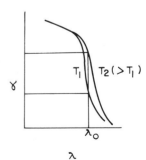

Figure 7.4. Temperature variation of the spectral response of the absorption coefficient (α) of a typical semiconductor.

conductivity at metal–insulator transition in oxides of V, Ti, Nb, Ni, Fe, and KrRb changes by several orders of magnitude within a narrow temperature range of about a degree. Since the metal–insulator transition is invariably accompanied by a structural transformation, this process is inherently slow and limits the response time of such detectors to about a second.

7.3. Thermal-Imaging Applications

Some of the above-described thermal detectors have been utilized in developing thermal-imaging systems. Whereas visual-spectrum images are formed by reflection and reflectivity differences, thermal images are produced by self-emission and emissivity differences. When operability at any time of the day or night and under all weather conditions is the primary consideration, thermal-imaging devices are superior to other electro-optic imaging devices. Because a thermal-imaging system makes it possible to see in the dark, it is also called a "night eye."

A thin film absorption-edge detector, consisting of a self-supporting thin film of Se backed by a thin film of Cr for absorption of radiation, was utilized by Harding et al.[23] in 1958 to make an image converter. The IR image of an object is focused onto the Se–Cr film (with the Se side facing the radiation) where it is stored as dark and light regions produced by local heating. The image is seen by transmission through the film using a sodium lamp. Recognizable images of a teapot at 60°C above ambient and of a man's head were obtained using this device. The response time of this device is approximately 0.5 sec.

Amorphous films of several chalcogenides of arsenic and germanium obtained by vacuum evaportation are known[24,25] to exhibit large photo-induced as well as thermally induced microstructural changes. These changes, some reversible and some irreversible, are accompanied by complex changes in the film thickness, optical constants, and the absorption edge. It is possible to utilize the thermally induced effects for the purpose of thermography.[26] The image of a hot object can be "printed" in the amorphous film and can be "read out" by passing light of appropriate wavelength through the film. Alternatively, the microstructural changes can be made visible by using suitable chemical etching techniques.[27]

Another thermal-imaging device is the pyroelectric vidicon.[28,29] In this system, the surface charge image created by IR radiation from a hot source is read out by a scanning electron beam, as in the case of an ordinary vidicon using photoconductive films (see Chapter 3), to produce an IR image of the hot object. Such night-vision devices are now commercially available.

7.4. Photothermal Conversion

As already mentioned, the heating effect of radiation can be used for direct conversion of solar radiation into heat[30] for innumerable applications involving heating or cooling, as, for example, in air conditioning of buildings and heating of water for numerous applications. Let us first discuss the requirements on the material for efficient photothermal conversion of solar energy into heat. Obviously, first of all it should be capable of strongly absorbing the dominant spectrum (0.3–2.0 μm) of solar radiation reaching the surface of the earth. This, however, is not a sufficient condition because as the absorber converts the intercepted solar radiation into heat, a part of this heat is lost by reradiation from its surface and thus significantly reduces the conversion efficiency. The heat loss can be minimized by minimizing the emittance and hence maximizing the reflectance (after Kirchhoff's law) of the material in the spectral region of reradiation from the converter surface. The position of the peak of the emission spectrum depends on the temperature of the converter and shifts to shorter wavelengths with increasing temperature (due to Wien's displacement law). For example, a terrestrial receiver kept at 650°C, an uppper temperature limit for most technological applications, radiates 95% of the total emission at wavelengths greater than 2 μm. An efficient photoconverter, therefore, should have high absorptance in the visible and the near IR, and low absorptance in the IR. Such a material is said to be spectrally selective.

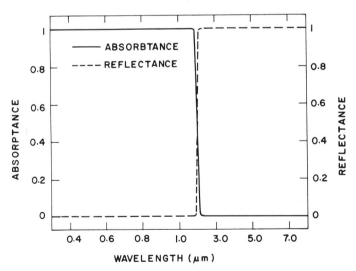

Figure 7.5. Spectral profile of the absorptance and reflectance for an ideal photothermal converter.

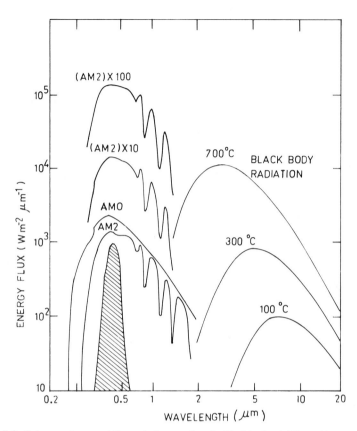

Figure 7.6. Solar spectrum and the emission spectra of a blackbody at different temperatures.

The spectral characteristics of an ideal spectrally selective converter operating under conditions of dominant reradiation and reflection losses are as shown in Fig. 7.5. Although the spectral ranges of solar input and emission loss are quite well separated, a small overlap is unavoidable, as shown in Fig. 7.6. Here, AM1 (air mass = 1) refers to the solar spectrum at sea level when the radiation falls normally on the surface. Similarly, AM2 refers to the solar spectrum at sea level when the radiation has traversed double the distance as in AM1, which occurs when the elevation of the sun is 30° with respect to the horizon. AM0 refers to the solar spectrum in the space beyond the earth's atmosphere. The spectral overlap causes loss of thermal energy in the part of the thermal emission spectrum which is trailing into the absorption region. This loss can be minimized by placing the cutoff wavelength in the spectral profile (Fig. 7.5) at the optimum location, determined by the temperature of the converter and the concentration of the solar flux.

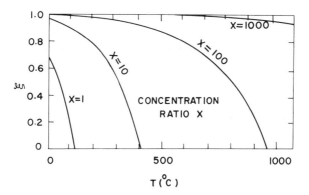

Figure 7.7. Effectiveness (ξ) of a blackbody absorber compared to an ideal selective absorber plotted as a function of absorber temperature T (after Ref. 31).

Let us now see under what conditions of operation a selective absorber provides really significant gains over the much simpler nonselective (for example, a blackbody) absorber. This is illustrated in Fig. 7.7 which is a plot of the effectiveness ξ of the nonselective blackbody absorber compared to an ideal selective absorber as a function of the absorber temperature and solar concentration. The quantity ξ is defined as the ratio of the energy absorbed and retained by a blackbody to that absorbed and retained by an ideal selective absorber. The figure shows that for a given solar concentration, the effectiveness of a blackbody decreases as the absorber temperature increases. For a given temperature, a selective absorber is essential unless the solar concentration is high. It is also clear from the figure that temperatures much in excess of 100°C are not possible without concentrators and the concentration required increases rapidly with increasing absorber temperature.

The above discussion shows that the absorptance α, defined as the fraction of incident solar flux absorbed by the absorber and mathematically given by

$$\alpha(T) = \frac{\int_0^\infty \alpha(\lambda)H(\lambda)\,d\lambda}{\int_0^\infty H(\lambda)\,d\lambda}$$

where $H(\lambda)$ denotes the incident radiation spectrum, and the emittance ε, defined as the fraction of radiant energy emitted by the absorber compared to that emitted by a blackbody (BB) at the same temperature and mathematically given by

$$\varepsilon(T) = \frac{\int_0^\infty \varepsilon(\lambda)\varepsilon_{BB}(\lambda, T)\,d\lambda}{\int_0^\infty \varepsilon_{BB}(\lambda, T)\,d\lambda}$$

are of key importance in determining the overall efficiency of a photo-thermal converter. The quantities α and ε in turn are determined by the optical properties of the material and the temperature and surface morphology of the converter surface. We now define a quantity α_m (called the *absorptance of merit*) which is a measure of the overall efficiency of conversion at a temperature T. This is given by

$$\alpha_m = \alpha - \varepsilon\sigma T^4/X\Phi$$

where σ, X, and Φ are, respectively, the Stefan–Boltzmann constant, the flux amplification, and the solar flux intensity. Thus α_m represents the balance between the actual absorptance α and the loss $(\varepsilon\sigma T^4/X\Phi)$ due to reradiation. Losses due to conduction and convection are neglected. Figure 7.8 shows the variation of α_m versus ε for an absorber with $\alpha = 0.90$ and 0.95, operating at a temperature of 500°C. This again shows that at low solar concentrations it is essential to have good selectivity.

In addition to the requirements of high α and low ε, the materials must also satisfy the following conditions: (1) long-term stability at the desired operating temperature, (2) stability against (or recovery from) short-term overheating due to failure to extract energy from the collector, (3) stability against atmospheric corrosion, (4) applicability to given susbstrate materials, (5) reproducibility, and (6) reasonable cost at a given operating temperature.

Selective coatings can be prepared by a variety of techniques such as electroplating, electroless deposition, chemical conversion, spray pyrolysis, vacuum evaporation, and, in particular, for paints by dip, spray, and electrostatic painting. The more specialized techniques of CVD, glow

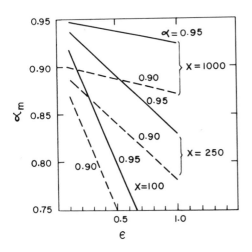

Figure 7.8. α_m as a function of ε for solar absorbers of $\alpha = 0.90$ and 0.95 and various flux concentrations X at an operating temperature of 500°C (after Ref. 30).

discharge, and sputtering are used for materials that are difficult to deposit by conventional techniques. The various types of selective coatings (summarized in Table 7.1) can be broadly classified as discussed in the following sections.

7.4.1. Metallic Surfaces

Because of their high reflectance in the thermal infrared and hence low emittance, metals seem attractive for photothermal applications. However, since they exhibit plasma edge at wavelengths much shorter than 2 μm (due to very high free-carrier concentration) they are poor absorbers in the visible and near IR. For example, noble metals with $\varepsilon = 0.02-0.04$ have $\alpha = 0.10-0.14$. Transition metals, on the other hand, have better absorber characteristics ($\alpha \sim 0.50$) but have high emittance. Polished Zn is an example of a natural surface with a fair degree of selectivity ($\alpha = 0.50$, $\varepsilon = 0.05$). A much higher degree of selectivity can be achieved when metals are used in conjunction with suitable semiconductors as described in the next section.

7.4.2. Metal–Semiconductor Tandems

The drawback of low absorptance of a metal surface can be overcome by depositing on its surface a semiconductor layer whose absorption edge lies in the vicinity of 2 μm. In such a tandem structure, the semiconductor layer absorbs the incident solar flux and, being transparent in the IR, allows the metal surface to see through it and thus supresses the thermal emittance. Selectivity is thus obtained due to high absorptance of the semiconductor and low emittance of the metal. A variety of such coatings using Si, Ge, and PbS as absorber materials and Al, Ni, Co, and Ag as reflector materials have been developed.[32-35] Because the refractive index of materials with absorption edge in the vicinity of 2 μm is high (greater than 3), an appreciable fraction of incident flux is lost due to high reflectance of the semiconductor surface. An AR coating at the semiconductor, therefore, becomes almost essential and SiO has been used mostly for this purpose. Recently highly stabilized amorphous Si films[36] prepared by CVD have attracted attention because of their much higher absorptance compared to crystalline materials. Use of transparent conducting oxides such as suitably doped In_2O_3 and SnO_2 in place of a metal reflector makes possible another tandem configuration in which the reflector lies at the top and the absorber lies at the bottom, for example, In_2O_3 on Si.[37] The oxide reflects the IR and the semiconductor absorbs the visible and near IR transmitted through the oxide. Figure 7.9 shows the spectral characteristic of such a coating. These

Table 7.1. Some Important Selective Solar Absorber Coatings

Material	Fabrication technique	α	ε (Low T)	ε (High T)	Stability (°C)
Al$_2$O$_3$	Anodization plus pigmentation	0.93	0.20		
ZnO	Anodization	0.95	0.08 (60)		
Steel	Acid dip	0.91	0.1 (100)		200
CuO$_x$	Spray pyrolysis	0.93	0.11 (80)		
Si (on Ag)	CVD + AR coating	0.80	0.05 (100)	0.07 (500)	500
Black Co (CoO$_x$)	Anodization	0.93	0.24 (260)		
Co$_3$O$_4$	Thermal oxidation	0.90	0.3 (140)		>1000
PbS	Vac. evaporation	0.98	0.2 (240)	0.3 (300)	300
	Paint	0.90	0.4 (100)		
Al$_2$O$_3$–Ni	Vacuum evaporation	0.94	0.10 (150)	0.35 (500)	
Al$_2$O$_3$–Pt	Vacuum evaporation	0.94	0.07 (150)	0.3 (500)	
Al$_2$O$_3$–Au	Sputtering	0.95	0.025 (20)		300
Al$_2$O$_3$–Cu	Sputtering	0.90	0.045 (20)		200
Black Cr (CrO$_x$–Cr)	Electrodeposition (on Ni)	0.95	0.07 (100)		350
	Electrodeposition (on Si)	0.97	0.06		
Black Ni (NiS–ZnS)	Electrodeposition	0.88	0.1 (100)	0.16 (300)	<220
	Electrodeposition, two-layer	0.96	0.07 (100)		<280
	Electrodeposition (on Zn)	0.94	0.09 (100)		<200
WC–Co	Plasma spray	0.95 (600)	0.28 (200)	0.4 (600)	>800
Al–PbS–SiO	Vacuum evaporation	0.89	0.018 (100)		
Ni–Ge–SiO	Vacuum evaporation	0.88	0.035 (100)		240
Ni–PbS–SiO	Vacuum evaporation	0.93	0.043 (100)		
Cr–Ge–SiO	Vacuum evaporation	0.93	0.11 (100)		240
Cr–PbS–SiO	Vacuum evaporation	0.94	0.12 (100)		240
Al$_2$O$_3$–Mo– Al$_2$O$_3$	Vacuum evaporation	0.85–0.95	0.34 (100)	0.41 (500)	>550
CoO + Fe$_2$O$_3$	Electroplating	0.90	0.07		
MgF$_2$–Mo– CeO$_2$–Mo	Vacuum evaporation	0.95	0.07		
MgF$_2$–Mo– MgF$_2$–Mo	Vacuum evaporation	0.89	0.075		
Ag–Al$_2$O$_3$ + Cr–Al$_2$O$_3$	CVD	0.90–0.95	0.02–0.04		
PbS–CdS	Solution growth	0.92	0.12		
PbS–ZnO (AR Coating)	Solution growth + spray pyrolysis	0.93	0.17		
WO$_2$ + Al$_2$O$_3$	RF sputtering	0.93	0.09		

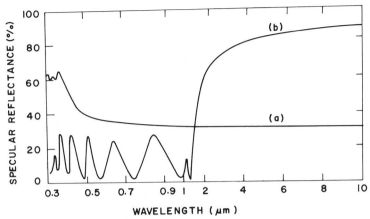

Figure 7.9. Spectral reflectance curves of (a) uncoated Si and (b) Si coated with 0.79 μm of indium tin oxide (after Ref. 37).

coatings are of great interest because the oxide coatings can be prepared by a very simple and inexpensive technique of spray pyrolysis.

7.4.3. Metal–Semiconductor Mixed Coatings

These coatings are single-layer coatings produced on a metallic base by electrodeposition techniques and are called *metal blacks*. These metal blacks are semiconductors which are absorbing in the visible and transparent in the IR (above 2–3 μm) and provide selectivity as in the case of metal-semiconductor tandems. Some of the black metals, for example, black Ni and black Co, can be prepared by the electroless technique also. The spectral characteristics of some of the black metals are shown[38–40] in Fig. 7.10. A number of chemically converted coatings on Cu, Steel, Al, and Zn are also available. The α and ε values of the various coatings are shown in Table 7.1. Very promising values of α and ε ($\alpha = 0.94$ and $\varepsilon = 0.06$) have been obtained[38] on electroless nickel in our laboratory. The optical and thermal properties of metal blacks are critically controlled by the plating parameters since they in turn control the composition of the coating. Black chrome is a mixed phase of Cr particles dispersed in the dielectric medium of Cr_2O_3. Nickel black is a mixed phase of sulfides and oxides of Ni and Zn with some free metal particles.

7.4.4. Interference Stacks

The well-established theories of all-dielectric and metal–dielectric multilayer optical coatings (see Chapter 2) can be used to produce a broad

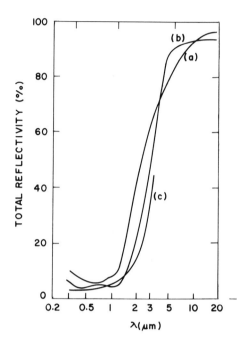

Figure 7.10. Spectral reflectance curves of (a) electroless Ni, (b) electroplated black Ni, and (c) electroplated black Cr (after Ref. 38–40).

transmission band in the solar spectrum region. Thus all the solar flux is passed onto the underlaying absorber layer without any reflection loss, and all wavelengths other than those in the solar flux are reflected back. AMA coating (Al_2O_3–Mo–Al_2O_3) is[41] a classic example of this type of structure. It is stable up to temperatures $\sim 800°C$ and is suitable for use in space applications. Interference stacks consisting[42] of alternate layers of solution-grown CdS and PbS have been fabricated by us. Figure 7.11 shows the calculated and measured reflectance of one such four-layer stack.

7.4.5. Particulate Coatings

Particulate coatings consist of absorbing metal or semiconductor particles dispersed in a dielectric medium and are based on the phenomena of reflective and resonant scattering. Reflective scattering is purely due to the geometry of the dispersed particles, while resonant scattering is determined both by the geometry and the optical properties of the particles and the surrounding medium. Chrome black is an example of particulate coating. A very interesting method[43] of fabricating coatings such as Al_2O_3–Ni consists of first making a highly porous coating of Al_2O_3 on Al by anodization and then embedding Ni particles in these pores by electrodeposition. Selective paints are another category of particulate coatings in which fine

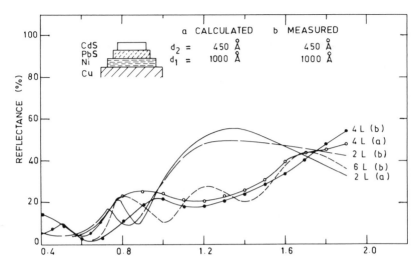

Figure 7.11. Calculated (a) and measured (b) spectral reflectance curves of 2,4,6-layer PbS–CdS interference stacks deposited on Ni–Cu.

particles of a semiconductor are mixed with a binder which is transparent in the IR and is of low refractive index, and the mixture is brushed or sprayed on a suitable substrate. A value of $\alpha = 0.9$ and $\varepsilon = 0.2$ to 0.3 can be obtained[44] for a PbS-in-silicone binder coating brushed onto Al.

7.4.6. Topological Coatings

Another way of obtaining selectivity on materials having intrinsically low values of emittance, for example, metals, is by modifying the topography of the surface. This is done by adjusting the surface roughness to dimensions of the order of the wavelength of visible light. This enhances the absorptance of the surface due to multiple reflections in the cavities formed. Since the roughness is of the order of the wavelength of visible light, it appears substantially flat to IR radiation. Nearly ideal surfaces of rhenium[45] and tungsten[46] have been produced by CVD. These surfaces possess dendritic structure which results in multiple reflections and hence high absorptance.

Another approach to changing the topography of a surface by providing holes of appropriate size by mesh–grid structures fabricated by lithographic techniques has also been suggested by some workers.[47,48] Advanced techniques of x-ray and electron beam lithography and use of holographically prepared gratings can make such structures feasible.

8

Surface Engineering Applications

8.1. *Introduction*

As the name implies, this chapter discusses the use of thin films for modification of engineering surfaces for protection of machine components and devices from environmental effects and mechanical damage, thus increasing their useful life and/or enhancing their aesthetic appeal. It is convenient to classify the various surface engineering applications into three main categories (1) surface passivation applications—those involving protection from environmental effects; (2) tribological applications—those involving protection from mechanical damage; and (3) decorative applications. Before going into the details of these applications, let us give a brief introduction of the subject of surfaces.

A *surface* may be defined as a two-dimensional interface between two media (or phases). The emergence[1-3] of a host of highly advanced surface analytical techniques for the structural and chemical identification of surfaces has contributed a lot to the basic understanding of surfaces. The results of some important studies may be summarized as follows. All surfaces, including those which appear smooth and are specularly reflecting, are rough on an atomic scale and have surface irregularities (asperities). Also, surfaces in general are not clean and have a variety of adsorbed impurities on them. The surface of a typical metal, for example, consists[4] of a thin (~ 0.01–0.1 μm), generally transparent layer of the corresponding oxide containing cracks and pores. Molecules of water, oxygen, and some polar substances, both organic as well as inorganic, are weakly bound to the oxide layer. Figure 8.1a shows a schematic picture of such a metal surface.

A variety of interactions are possible at an interface between two media. These surface interaction properties are not fundamental or intrinsic in the same sense as Young's modulus or thermal conductivity, but are determined by a number of external parameters such as cleanliness of the surfaces, environment, whether or not the surfaces are in relative motion, and, of course, the surface properties. Among the surface properties, those

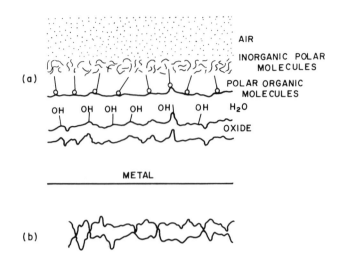

Figure 8.1. (a) "Hierarchy" of spontaneously adsorbed layers on a metal surface; (b) schematic diagram showing contact between two solid surfaces.

which are of great importance in determining the interfacial interactions are: (1) chemical reactivity of the surface, i.e., the tendency of the surface to acquire a surface film of composition different from that of the bulk, (2) tendency of the surface to adsorb impurities from the environment, (3) the surface energy which determines the work done to create a fresh surface, and (4) the interfacial energy which determines the compatibility of the two interacting surfaces. Mechanical properties come into picture when the surfaces are in relative motion. When the surfaces are moving at very high speeds relative to each other, the temperature dependence of the various properties also becomes important.

The various surface interactions, either singly or in combination, lead to damage of the surface of devices such as cutting tools, mechanical components, waterpipes, etc. An ideal solution to this problem of damage would be to use a single material for device fabrication, which in addition to satisfying the required mechanical, electrical, and thermal properties, is simultaneously resistant to the related chemical environment. However, such a solution is not practically achievable in general because of cost or weight considerations in some cases and mutually incompatible requirements in others. The use of surface treatments and thin surface films to achieve the desired characteristics provides an elegant solution to the problem of combating surface damage and hence increasing the service capabilities and reliability of devices and reducing losses of materials. The advantage of using thin films for protection and decoration lies in the possibility of manufacturing engineering components from cheap, readily available materials. It is not necessary, for example, to build the whole waterpipe using special expensive corrosion-resistant alloys; a thin coating of the corrosion-resistant material on the inner surface of ordinary water-pipe is sufficient for corrosion prevention. Thin films have found application in a large number of engineering systems such as automobiles, aircraft engines, turbine blades, ball bearings, machine tools, ultrahigh vacuum components, high-temperature furnaces, ship components, bridges, water-pipes, nuclear reactors, etc.

The general problem of protecting materials consists of two major parts: (1) the search for a material having the desired characteristics and (2) the development of a suitable technique for deposition of this material. Since the various deposition techniques have already been discussed in Chapter 1, only a brief mention will be made of the relevant techniques here.

8.2. Surface Passivation Applications

This section deals with applications of thin films to make a surface passive[5-9] and chemically inert with respect to its environment. For example, one may require an oxidation-resistant metal surface especially at high temperatures, as for high-temperature furnaces, or a passivated surface between the material of a container and its contents, or between an electrode and the electrolyte. The need for passivation arises because the interaction, if any, between the surface of the material and its chemical environment leads to deterioration of the material. This kind of gradual deterioration of materials arising due to their chemical reaction with the environment (gaseous or liquid) is called *corrosion*. Corrosion may occur either as a direct chemical or an electrochemical attack. Conditions favoring the former type of attack are more often provided by contact with gases or moisture-containing atmospheres, while immersion in liquids favors the latter type of attack. This, however, is by no means a generalized rule. A corrosion reaction, although it starts at the surface, may proceed further down to the bulk if the reaction-product layer formed is porous. Therefore, a corrosion-resistant coating in addition to being inert should also be continuous and nonporous.

Engineering materials which require protection from corrosion include iron, steels, copper, nickel, cobalt, chromium, graphite, carbides, borides, nitrides, etc. During corrosion these materials undergo oxidation and, depending on the environment and the thermodynamic conditions, a hydroxide, oxide, carbonate, sulfide, nitride, or halide may be formed as the reaction product. Of the various methods of corrosion prevention, the use of protective coatings is of considerable interest. Coatings provide protection by acting as barriers between the object and its environment. A proper choice of the coating material has to be made depending on the environment in which the material is to be used. The various corrosion-resistant protective coatings can be classified into the following categories: (1) coatings of reaction product, (2) metallic coatings, (3) inorganic coatings, and (4) organic coatings. The coating material and the coating technique selected for protection depends on such factors as the size and shape of the object, material of the object, the appearance requirements, and the environment in which it is going to be used. For example, small articles are generally metal coated; structural steel, such as that used in construction of automobile bodies and bridges, is always painted. Inorganic coatings are used for protection against highly aggressive environments such as those found in chemical industries involving contact with corrosive liquids or exposure to high temperatures.

8.2.1. Coatings of Reaction Product

The natural oxide film formed spontaneously on many metals offers resistance to further chemical reaction if it is nonporous and adherent. Also the oxide film should be readily renewable by immediate oxidation of the metal if during service it is accidentally removed. These conditions are fulfilled to a large extent by Al_2O_3 films on Al and its alloys. This makes possible the use of Al and its alloys in the aircaraft industry. For resistance in some highly aggressive environments requiring thicker coat-ings, Al_2O_3 on Al is grown by anodization. Similar protection to Cr and Ni–Cr steels is provided by thin films of Cr_2O_3 or $FeO \cdot Cr_2O_3$. These films provide protection not only against atmospheric oxidation but also against aggressive corrosive attack.

8.2.2. Metallic Coatings

The use of metallic coatings for protection of corrodable materials such as Fe and steel has been known for long. The most common metals used for protection of Fe and steel are Zn, Ni, Cr, Sn, Cd, Pb, and Al. Depending upon the size of the object to be protected and the protecting material to be used, metal coatings may be applied by any of the following methods: (1) dipping the object in molten metal, (2) spraying, (3) electro-plating, and (4) cementation. Cathodic protection is provided by metals which are anodic with respect to the base metal, for example, Cd and Zn on Fe. In the case of coating metals which are cathodic with respect to the base metal, protection depends solely on physical coherence and the non-porous nature of the films. Zinc is widely used for protection of steel sheet and wires exposed to the atmosphere. Nickel coatings, generally electroplated, are used for protection of household appliances, equipment for dairy and food industries, plumbing fixtures, the inner side of steel pipes, automobile parts such as bumpers, etc. Chromium coatings are also generally applied by electroplating and, in addition to corrosion resistance, also enhance the appearance of the object. Tin coatings are applied to the inner side of steel cans used for packaging food and beverages, thus protecting them from corrosion. Lead coatings are used for protection of surfaces exposed to sulfuric acid. Aluminum thin films are used for the protection of structural steel. Cadmium coatings are applied for the protection of products intended for indoor use where a bright appearance is desired.

Coatings produced by cementation (or diffusion) are alloy–intermetallic compound coatings and their formation depends upon the diffusion and solubility of the coating metal on the base metal. Depending on the diffusing element, the finishing process is called chromizing, siliconizing, aluminizing, etc.

The chromized surfaces of iron and carbon steel show resistance to alkaline salt and oxidizing acid solutions. They also resist atmospheric oxidation for long periods of time up to a temperature of 800°C. For temperatures above 800°C, interdiffusion of Cr into the substrate becomes significant, thus reducing the Cr content at the surface and hence the resistance to corrosion. The resistance to oxidation at higher temperatures can be increased using Cr–Al–Fe and Cr–Si–Al–Fe mixed coatings which can withstand atmospheric oxidation up to 1100°C.

Aluminized steel is highly resistant to atmospheric oxidation and sulfurous gases at high temperatures and can withstand a temperature of 750°C for years. Aluminized Ni and Cr-based superalloys can resist oxidation and sulfurous-gas attack up to a temperature of 1200°C. Aluminized steel finds use in the oil-refining industry for tube stills for cracking and topping, valves, retorts, and condenser parts. Steel strip coated with Al is used in the packaging industry.

Aluminides of Nb, Ta, and V, silicides of Mo, W, Nb, Ta, and V, and berrylides of Nb and Ta formed by diffusion of Al, Si, or Be into the refractory metals (Na, Ta, Mo, W, V) protect them from oxidation. Refractory metals are of great importance in gas turbines and space vehicles.

8.2.3. *Inorganic Coatings*

Oxides of materials such as Al, Zr, Ti, Si, and their mixtures are known for their low porosity and excellent adhesion to the base material and are therefore used for surface passivation applications. These coatings are also heat and thermal-shock resistant. Al_2O_3 coatings provide protection against oxidizing and reducing atmospheres up to a temperature of 1500°C. ZrO_2

is a very good heat-resistant material and is used in nuclear reactors. The heat-resistance property along with the catalytic action of ceric oxide in the combustion process makes it useful for coating on the inner surface of combustion chambers. An addition of 3% TiO_2 improves the physical properties of ceric oxide but reduces the maximum temperature it can withstand against oxidizing atmospheres from 1920°C to 1100°C. Metals such as Ni and Co can be added to these oxides to make them resistant to higher temperatures.

Another important class of inorganic coatings used for protection of Fe and steel consists of phosphates of Fe, Zn, and Mn, produced by immersing in or spraying of hot solutions of the corresponding phosphates or their mixtures with an excess of phosphoric acid. It should be emphasized that phosphated surfaces by themselves do not have very good protecting properties because of their porous nature but provide an excellent base for organic protective coatings, which are discussed below.

8.2.4. Organic Coatings

A large number of organic materials such as paints, lacquers, and enamels are used for protecting metal surfaces from atmospheric corrosion. The coatings can be applied by painting, dipping, or spraying, depending on the size and shape of the object to be coated. They provide protection by acting as mechanical barriers between the object and its surroundings. Prior to organic coating, the surface of the base metal is phosphated to improve the adhesion of the coating. These types of coatings are used on a variety of items such as refrigerators, steel almirahs, toys, automobile bodies, etc. In addition to protection, they also add to the aesthetic value of the finished products.

8.3. Tribological Applications

Tribology is the science and technology of interacting surfaces in relative motion.[10-21] Mechanical damage of either or both the surfaces occurs in general whenever two surfaces are rubbing against each other in a device. A large number of components of engines, machines, motors, etc. undergo constant rubbing and therefore wear out after a certain period of use. Thin films of mechanically strong materials on the surface of such components are used to increase their service life. Another problem which occurs when surfaces are in relative motion is that of friction. Thin films of certain materials between two moving surfaces can be used to provide lubrication and hence reduce friction. Thus, based on the function they serve, tribological coatings may be classified as: (1) wear-resistant coatings and (2) lubricating coatings.

8.3.1. Wear-Resistant Coatings

Wear[22] may be defined as undesirable and unintentional deterioration of objects due to use by the removal of material from one or both of the rubbing surfaces. Although this definition of wear sounds very simple, it encompasses in itself a number of diverse phenomena which may operate singly or in combination to produce wear. These phenomena are listed below.

(1) *Adhesive Wear.* Since the surfaces in general are rough, contact between them, except at very high pressures, takes place at asperity tips only, as shown in Fig. 8.1(b). When two metal surfaces are rubbing against each other, intense local heating and high pressures are produced at the interface by colliding asperities. This results in the formation of metallic junctions (the process being also known as *cold welding*) which are of appreciable size by molecular standards. This is only the first stage of the adhesive wear mechanism and does not directly lead to loss of any material from the system. During shearing of these microwelds material is plucked from the softer of the two surfaces and transferred to the harder one, thus giving rise to adhesive wear, also called scoring, scuffing, or galling. Further complications arise if some secondary mechanism encourages the transferred particle to break away. These broken particles, which are generally strengthened by work-hardening at the welded regions, then cause further wear by abrasion as discussed in the following paragraphs.

(2) *Abrasive/Cutting Wear.* In this type of wear, removal of the material from a surface is accomplished not by its sticking to the other surface but instead by its being ploughed or gouged out by a much harder surface. The harder surface can be either one of the two rubbing surfaces or a third body, for example, a small particle of grit caught between the two rubbing surfaces. Abrasive wear occurs when the two rubbing surfaces differ widely in their hardness and the harder surface has a certain degree of roughness, so that its surface asperities dig into the softer surface, literally ploughing out furrows which break away forming loose wear particles.

This type of wear can be reduced by increasing the hardness at the surface, either by some surface treatment or by applying overcoatings of hard materials using thin film technology. Hardening of the material throughout the volume has the disadvantage of decreasing the overall durability and also the surface hardness to relatively low values. The concept of hardness is related to the idea of solidity and compactness and represents the resistance of a material to indentation, scratch, abrasion, cutting, or drilling.

The resistance of a surface to abrasive wear not only depends on its hardness H but also on the elastic modulus E of the surface material. The ratio H/E represents the elastic limit of strain, that is, the amount of elastic deformation that the surface can sustain, and gives the true measure of abrasive-wear resistance. Thus when an abrasive particle comes in contact with the surface in question, the surface would deform elastically to get out of the way of the particle and would return elastically to the original configuration without getting abraded when the particle has passed.

(3) *Chemical Wear.* This type of wear occurs when a surface prone to corrosion is being rubbed against another surface. This happens when the reaction-product layer formed by corrosion is not adherent and wear-resistant and can be easily removed by rubbing, thus exposing the surface of the material to the environment and allowing the chemical attack to continue further. The loose debris formed by particles of the corrosion product, which are generally harder than the surface, can cause further wear by abrasion.

(4) *Surface Fatigue Wear.* Fatigue wear occurs at surfaces in rolling contact and is characterized by sudden pitting or flaking-off of the surfaces on a large scale due to repeated cyclic stresses, usually terminating the life of the mechanical system.

A large number of surface treatments and coatings are available for combating wear. Case-hardening is a metal-treatment process which produces a hard surface, called the case, leaving the bulk of the metal relatively soft, thus resulting in a unique composite property, that is, a hard, wear-resistant case covering a strong and tough core. The main case-hardening processes for steel are carburizing, nitriding, carbonitriding, siliconizing, and chromizing and involve diffusion of the corresponding element into steel either by thermal means or by ion implantation.

Wear-resistant coatings are coatings of hard materials such as Cr, Co, Si, Mo, W, C, and their alloys, and carbides and nitrides of refractory metals and their mixtures along with their composites with metals. The mechanical properties of a large number of coatings of pure metals, alloys, oxides, carbides, and nitrides prepared by vacuum evaporation and sputtering processes have been reviewed by a number of workers.[23-25] Other techniques which are in common use to produce wear-resistant coatings are electroplating, electroless plating, electrophoretic plating, and CVD. The properties of the coatings may depend on the techniques of deposition and the deposition parameters, as illustrated in Figs. 8.2–8.4. Figure 8.2 shows[26] the dependence of the hardness of TiC coatings prepared by low-pressure plasma deposition (LPPD) on the pressure of the reactant gas (acetylene) for (1) a fixed substrate and (2) a rotating substrate. Figure 8.3 shows the results of machining tests carried out on uncoated and coated tips prepared by activated reactive evaporation (ARE), LPPD, and CVD.

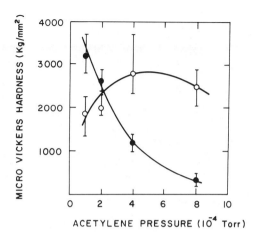

Figure 8.2. The Vickers microhardness of TiC films prepared by LPPD as a function of acetylene pressure for a fixed substrate (○) and a rotated substrate (●) (after Ref. 26).

The figures show the crater depth and the flank wear data after lathe machining of a Ni–Cr–Mo steel SNCM-8 rod (corresponding to SAE 4340) at a feed rate of 0.4 mm rev^{-1} with a depth of cut of 0.5 mm and a machining speed of 234–158 m min^{-1} for 0–15 min and 254–141 for 15–23 min. Figure 8.4 shows[26] scanning-electron micrographs of the rake and flank faces of the uncoated and the TiC-coated tips after 15 min running. Hardness and

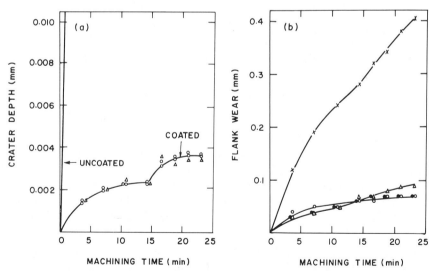

Figure 8.3. The crater depth (a) and flank wear (b) as a function of machining time for coated and uncoated tool tips [SNCM8, corresponding to SAE 4340: feed, 0.4 mm per revolution; depth, 0.5 mm; speed, 234–158 m min^{-1} (750 rpm) for 0–15 min and 254–141 m min^{-1} (1200 rpm) for 15–23 min]; (×) uncoated; (○) TiC-coated (activated reactive evaporation); (●) TiC-coated (low-pressure plasma deposition); (△) TiN–TiC-coated (CVD) (after Ref. 26).

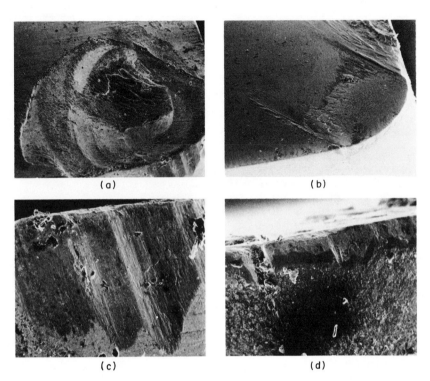

(a)   (b)

(c)   (d)

Figure 8.4. Scanning-electron micrographs of the faces of TiC-coated and uncoated tips after 15-min running: (a) rake face of the uncoated tip; (b) rake face of a coated tip; (c) flank face of the uncoated tip; (d) flank face of a coated tip (after Ref. 26).

other physical characteristics such as adhesion and the coherent nature of electroplated Cr make it useful for coating on cutting tools, files, dies, rolls for papermaking machinery, etc. Al-alloy pistons of internal combustion engines are Sn-plated to prevent wear of the cylinder walls by the abrasive Al_2O_3 particles during the running-in period. Electrotypes are plated with Cr or Ni for longer press runs. Electrophoretic deposition of WC–Co composite followed by sintering has been used[27] for protection of cutting tools. This composite material has the advantage of increased mechanical strength and toughness against premature fracture due to high stresses generated during use. This is so because when a crack formed within a hard region meets the soft Co matrix, it is arrested without extending further because the soft region can deform easily and thus relax the local stresses in the film. Figure 8.5 shows[28] the relative performance of different materials when sputter-deposited on cutting tools.

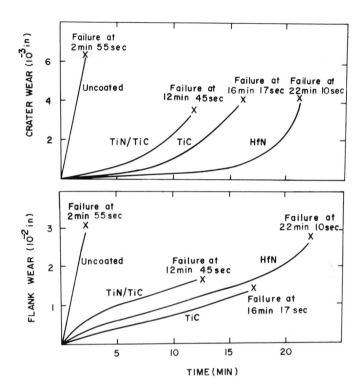

Figure 8.5. Crater wear depth (a) and flank wear depth (b) as a function of cutting time for various cutting tools (AISI 4340 Rc 36: speed, 700 ft min^{-1}; feed, 0.0052 in. per revolution; depth, 0.05 in.) (after Ref. 28).

8.3.2. Lubricating Coatings

As the name implies, these coatings are used to provide lubrication between two surfaces rubbing against each other and thus reduce friction. Friction is the resistance to motion which exists when two solid surfaces slide over each other. The main source of frictional forces arises from the shearing of the junctions (cold welds) formed at the real regions of contact when two clean surfaces are brought together. Moreover, if one of the sliding surfaces is much harder compared to the other, it penetrates into the softer surface and ploughs out grooves in the softer one. The total force of friction is thus the sum of the force required to shear the junctions and the force required to displace the material in front of the moving harder surface.

The most effective way of reducing friction is to prevent contact between the two surfaces by inserting between them a thin film, called the lubricating layer, which is much easier to shear than the underlying solid. For example, a thin film of a soft material such as In on tool steel reduces its friction by a factor of 10. Figure 8.6 shows[29] the effect of film thickness on the frictional properties of thin films of In on tool steel. When a hard metal surface is covered by a thin film of softer material, the surface area of the contact is determined by the hardness of the underlying metal, but shearing occurs in the softer metal layer. Other soft materials which can be used in thin film form for lubrication purposes are Au, Ag, Sn, Pb, Cd, PbO, CaF_2, etc. Such films gradually wear away and need to be replenished.

Another approach to reducing friction is to coat the rubbing surfaces with lamellar materials like graphite, MoS_2, WS_2, $NbSe_2$, $MoSe_2$. These materials are strong in compression but very weak in shear due to intercrystalline slip. One of the major difficulties with these substances is achieving strong adhesion to the substrate, because the cleavage planes are low-energy surfaces. Sputtered MoS_2 films, generally, have strong adhesion to metallic, ceramic, glass, and polymer surfaces, and therefore films ~ 2000 Å are thick enough to provide effective lubrication. A low coefficient of friction (~ 0.04) and long endurance lives (over a million cycles) have been obtained for these films. Surfaces of Cu, Ag, and bronze require oxidation prior to MoS_2 deposition for obtaining good adhesion. A low-friction element can also be prepared by electrophoretically depositing a reducible metal oxide or sulfide with nonreducible lubricating MoS_2 followed by H_2 reduction to produce a metallic matrix incorporating MoS_2.

As we have already mentioned, phosphated surfaces are porous in nature. These can be used in conjunction with lubricating oils to reduce

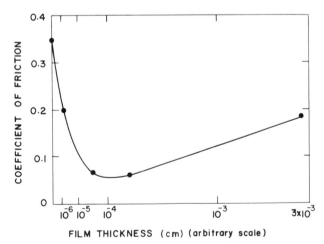

FILM THICKNESS (cm) (arbitrary scale)

Figure 8.6. Effect of film thickness on frictional properties of thin films of In on tool steel (after Ref. 29).

friction. The coatings, being porous, can adsorb large quantities of oil and thus act as reservoirs of lubricating oil; hence they are beneficial during break-in periods. In industry a variety of parts such as piston rings, cylinder liners, rocker arms, gears, bearings, shafts, nuts, bolts, refrigerator and automobile bodies, cartridge clips, etc. are coated with phosphates of Mn and Zn. Electroplated Cr can also be made porous under special conditions and be made to hold oil efficiently. These coatings are used as cylinder liners for internal combustion engines and on ball bearings.

Yet another approach to reducing friction is the use of hard wear-resistant coatings. Very little deformation takes place on the surface of such hard coatings and generally a thin film of moisture or adsorbed oxygen is sufficient to provide lubrication. Spalvis[30] has found that a composite coating of a hard wear-resistant film and a soft lubricating film has an increased endurance life. Sliney[31] has developed multicomponent lubricating coatings for use over wide temperature ranges. These coatings are mixtures of nichrome, CaF_2, glass, Ag, and oxides of Ba, Ca, K, and Si. The coatings were obtained by the plasma spray process and were found to be self-lubricating in the cryogenic to 870°C temperature range. Friction coefficients have been found to depend on microstructure and composition.

Some organic polymers such as PTFE and polyimides, fats, soaps, and waxes have also been sputter-deposited as thin lubricating films.

8.4. Decorative Applications

The use of thin films for decorative purposes[32,33] is a very old art. An almost endless list of articles exists which are electroplated for improving their aesthetic appeal. Some of the common ones are automobile and aircraft parts, refrigerator hardware, electrical appliances, office furniture, razor blades, photographic equipment, firearms, handbag frames, pens, costume jewelry, etc. Usually these decorative films provide protection also. Common metals which are electroplated for decorative purposes are Cr, Ni, Cd, bronze, Ag, Au, brass, Cu, and Rh. Interference effects (discussed in Chapter 2) in thin films can give rise to beautiful and pleasing color effects. Thin films of oxides, transparent sulfides, and fluorides are used for this purpose on glass bangles, lamp shades, and a number of other decorative articles made of glass.

A large number of articles made of aluminum such as jewelry, decoration pieces, toys, electronic instrument bodies, etc., are also beautified by anodization (Chapter 1). Aluminum surfaces can be easily anodized in a porous form which has high adsorption characteristics. Dyes of any color or metallic microparticles can then be incorporated into these films to give the desired appearance.

8.5. *Miscellaneous Applications*

This section describes some of the miscellaneous surface engineering applications which do not fall under any of the classified headings.

8.5.1. *Adhesion-Promoting Coatings*

The use of phosphate coatings for improving the adherence of organic coatings to metals has already been mentioned. In this procedure, adhesion is improved because of the porous nature of the phosphate coating which provides mechanical interlocking of the organic film with the base substrate. Similarly, use of thin layers of Cr and Ti for improving the adherence of gold contacts to ceramic substrate is well known. Among the recent developments is the use of certain chemicals known as coupling agents or adhesion promoters for improving the adhesion between polymeric resins and substrates such as glass, metal oxides, Si, GaAs, etc. Prior to deposition of the desired film, a thin film of these chemicals, for example silanes, is deposited on the substrate either by dipping in a solution of the resin, or by vacuum evaporation. The latter method[34,35] is particularly advantageous in cases where the substrate is susceptible to corrosion in a liquid medium. In contrast to the phosphate type of coatings, the overall adherence of the organic coating to the substrate is improved due to increase in the basic or fundamental adhesion between the two materials through interfacial molecular interactions.

It has been established[36,37] that the most important of these interactions are either Van der Waals type or acid–base (that is, electron acceptor-donor) in character. For a detailed discussion of the mechanism of adhesion promotion, the reader is referred to the published literature.[38–40]

8.5.2. *Preparation of Heterogeneous Catalysts*

Heterogeneous catalysts[41] are solids which increase the rate of thermodynamically feasible chemical reactions by virtue of certain specific properties of their surfaces. The activity of a catalyst resides chiefly at its surface which must be capable of adsorbing at least one of the reactants. This means that efficient catalytic surfaces must have very large surface areas, which can be obtained as finely divided and/or porous structures as those found, for example, in colloids, gels, sponges, microcrystalline powders, and thin films. Thin porous films may have real surface areas hundreds of times the geometrical surface area. Of the various methods of obtaining large-area surfaces, thin film techniques, especially the PVD techniques, provide an ideal method for preparation of catalytic surfaces from the point of view of both reproducibility and cleanliness of the surface. It should be noted that preparation of a microcrystalline powder, colloid, or gel in a form free from contaminations is extremely difficult, if not impossible. These preparations invariably contain traces of the parent compounds, which may drastically affect the activity of the catalyst.

Heterogeneous catalysts may be classified as metallic and nonmetallic (generally oxides and sulfides). Metallic catalysts are mainly used for hydrogenation, dehydrogenation, and hydrogenolysis reactions. Of the nonmetallic catalysts, the semiconducting ones are mainly used for oxidation, reduction, dehydrogenation, and cyclization reactions, and the insulators for dehydration and isomerization reactions. Salts and acids are used for polymerization, isomerization, cracking, alkylation, and hydrogen-transfer reactions. Most of the studies of thin films for catalytic purposes have been performed on metals only. The use of vacuum-evaporated metal films as catalysts dates back to the work of Beeck and his associates[42] in 1941. They studied the thin films of a number of metals and observed some correlation between the electronic structure of the metal and its catalytic activity. It was found, for example, that the activity of Rh thin films for hydrogenation of ethylene to ethane was at least 1000 times higher than that of Ni films. Catalytic activity also depends on the orientation of the film; for example, an oriented Ni film is about five times more active than an unoriented film. Beeck and his associates also found that the activity of Ni films for the hydrogenation of ethylene increases manyfold by the addition of 20–50 at.% Cu. However, for the hydrogenation of styrene, the catalytic activity of Ni films is drastically reduced even when small amounts of Cu are added. Electrodeposited films[43] of suitable metals have also been found to be active catalysts, but they are highly susceptible to poisoning by Hg and tap grease. With advances in surface analytical tools for studying various processes occurring at catalytic surfaces, the field of thin films will definitely contribute more to the technology of heterogeneous-catalyst preparation.

8.5.3. *Preparation of Nuclear Fuels*

One of the important types of nuclear fuels[44] consists of particles of fissionable material, with each fuel particle encapsulated in a thin film covering which serves any or all of the following purposes: (1) to protect the fuel from corrosion and embrittlement due to the matrix material or the coolant, (2) to protect the fuel from hydrolysis by moisture in the air in the case of fuels containing thorium or plutonium carbides, (3) to protect the matrix of a dispersion fuel from fission-recoil damage, (4) to minimize diffusion and decomposition of the fuel at the high temperatures prevailing during irradiation, (5) to retain fission products, thus reducing the shielding and maintenance costs, (6) to minimize the hazards associated with plutonium fuels, and (7) to provide fabrication of dispersions at lower costs and with better fuel distribution.

CVD techniques are generally used for coating fuel particles. The earliest experiments on coating of fuel particles were performed[45] at the Argonne National Laboratory of the U.S. Atomic Energy Commission in 1949, when uranium spheres were coated with electroless Ni to prevent recoil damage to the cladding in a fast reactor. Electrodeposited films of Cu were then developed[46] for use on berrylium to protect it against a sodium environment. A number of CVD methods are available[47–49] at present for coating oxides (as that of Al and Be), carbon, carbides, and their mixtures on a variety of fuel substrates in both porous and nonporous form. Porous structures serve to provide space for irradiation-induced expansion of fuel and accumulation of fission-product gases. The porous regions also serve as barriers to crack propagation.

8.5.4. Fabrication of Structural Forms

Besides modification of surfaces, thin film technology allows fabrication of a variety of unsupported structures[50] such as sheets, tubings, rings, dies, molds, rocket nozzles, etc. This technique of fabrication of objects is called *vapor forming*. Most of the work done in this field is based on CVD processes. A special case of vapor forming used for the fabrication of metallic objects is electroforming which uses electroplating as the deposition technique. Basically, in vapor forming the material is deposited on a mandrel to form the desired structure. The mandrel can be either temporary or permanent and accordingly removed from the deposit by appropriate means such as melting or dissolving. In some special cases, as for example, pyrolytic graphite, because of its unique properties[51] the mandrel can be made to fracture and fall away from the deposit during deposition. Alternatively, it can be made to shrink away from the deposit while cooling.

Credit for the idea of direct formation of structures by vapor deposition goes to Mond[52] who in 1891 suggested the fabrication of Ni tubes, sheets, and other structures by pyrolytic decomposition of nickel carbonyl on graphite cores. This process was commercially used for making dies and molds for a long time but was later abandoned in favor of the commercially well-established electroforming process[53] for Ni from the sulfamate–nickel bath because of the highly toxic nature of nickel carbonyl.

Van Liempt[54] in 1932 prepared refractory metal tubes by depositing W on a core of Mo or Cu by thermal decomposition of WCl_6 vapors at 2000°C on Mo and by H_2 reduction of WCl_6 on Cu and then dissolving out the Mo or melting out the Cu core. Other significant materials studied for vapor forming using CVD processes are carbides and boron nitride.

A large number of miniaturized mechanical structures such as cantilevers, plates, stretched membranes, etc. for use in microelectronic devices to function as vibration sensors, light valves, and tuning devices can also be fabricated[55] using the combined techniques of thin film deposition, photolithography, and etching.

8.5.5. Biomedical Applications

This section briefly discusses some of the applications of thin films in biomedical engineering. Because of lack of space it is not possible to go into details of each one.

Glass metal-clad microelectrodes developed[56] by sputtering and vacuum-evaporation techniques overcome many of the limitations of the conventional fluid-filled glass capillary microelectrodes for use in neurophysiological investigations. These thin film microelectrodes have the important property of low resistance and long shelf life. Thin film technology thus offers the possibility of a batch process for preparation of microelectrodes with uniform and reproducible characteristics and the development of other new types of specialized structures. Figure 8.7a shows a schematic diagram of the simplest kind of glass metal-clad microelectrode.

The potential of thin film technology along with photolithography has also been realized[57,58] for fabricating arrays of microelectrodes compatible in dimensions with those of nerve fibers ($\sim$ a few microns), for use in biorecording and biostimulation. Figure 8.7b shows a schematic diagram of a four-electrode probe for biostimulation.

Another application in this field is for fabrication of insulated electrocardiographic (ECG) thin film electrodes. Insulated ECG electrodes are superior to the conventional conductive electrodes[59,60] employing some type of a metal–electrolyte system because the latter cease to be effective when long-time measurements are required or when the subject is in motion. Further the electrolyte dries out with time and therefore needs to be replenished. It may also cause irritation and infection of the skin with which it is in contact after long periods. Insulated ECG electrodes have two more advantages: (1) they can be applied directly to the unprepared skin, and (2) dc drift is blocked directly at the electrode–skin interface. These insulated ECG electrodes are prepared by RF sputtering of dielectrics such as Ta_2O_5, TiO_2, $BaTiO_3$, and SiO_2 on Si substrates and are far superior in performance to those prepared by paste techniques.

Recently, Barth and Angell[61] have reported the fabrication of thin film linear thermometer arrays for the measurement of one-dimensional temperature profiles in tissues during cancer treatment by localized hyperthermia. Each array is comprised of six discrete Si diodes deposited on three flexible stainless steel wires with a maximum cross-sectional

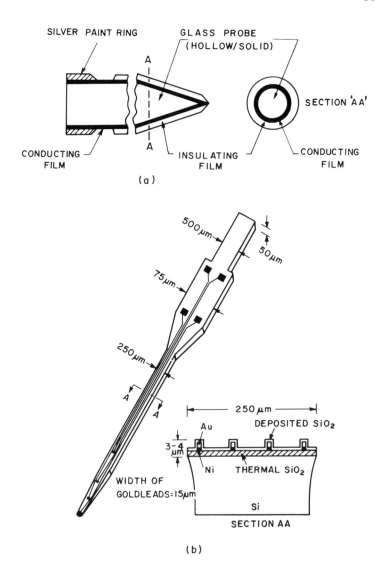

Figure 8.7. Schematic diagrams of (a) thin film metallized glass microelectrode and (b) four-electrode probe for biostimulation (after Ref. 57).

dimension of 0.5 mm, so that it is possible to introduce it into the tumor area through a small puncture wound.

Another recently reported[62] interesting application of thin film technology in the field of biomedical engineering is a thin film multielectrode array for cochlear prosthesis. The electrode array consists of photo-lithographically defined Pt on Ta conductors sandwiched between two polyimide layers. The Ta layer serves to improve the adhesion of Pt to polyimide. The array is designed to be inserted into the spiral scala tympani chamber through the round window of the cochlea. The successful fabrication of this array offers an important alternative to the conventional bundle electrodes.

Glow-discharge plasma-polymerized (GDPP) films[63] of certain polymers have been investigated by some workers and are seriously being considered for a number of biomedical applications. For example, extremely adherent GDPP thin films of propylene have been found[64] to be very promising for the protection of microelectrodes of oxygen sensors planted inside the human body from the poisoning effects of the surrounding aqueous environment containing ions and protein macromolecules. Besides this, GDPP films of methane, tetrafluoroethylene (TFE), and tetramethyldisiloxane (TMDSiO) on Pt electrodes and feedthrough wires also improve[65] the bonding strength of the platinum surface to the surrounding epoxy insulation, thus protecting the insulation from being disrupted by the biological environment and increasing the functional life of internally implanted electronic packages, such as oxygen sensors, cardiac pacemakers, etc. Some of the GDPP films have also been found to be biocompatible.[66] For further details on biomedical applications of GDPP thin films, see Refs. 67–72 of this chapter.

References

References for Chapter 1

1. J. L. Vossen and W. Kern, *Phys. Today* **33**, 26 (May 1980).
2. J. M. Poate and King-Ning Tu, *Phys. Today* **33**, 34 (May 1980).
3. J. P. Hirth and K. L. Moazed, in *Physics of Thin Films*, Vol. 4, G. Hass and R. E. Thun, eds., Academic Press, New York (1967), p. 97.
4. K. H. Behrndt, in *Physics of Thin Films*, Vol. 3, G. Hass and R. E. Thun, eds., Academic Press, New York (1966), p. 1.
5. R. E. Thun, in *Physics of Thin Films*, Vol. 1, G. Hass, ed., Academic Press, New York (1963), p. 187.
6. K. N. Tu and S. S. Lau, in *Thin Films—Interdiffusion and Reactions*, Wiley, New York (1978), p. 81.
7. L. Holland, *Vacuum Deposition of Thin Films*, Wiley, New York (1956).
8. K. L. Chopra, *Thin Film Phenomena*, McGraw Hill, New York (1969).
9. L. I. Maissel and R. Glang, eds., *Handbook of Thin Film Technology*, McGraw-Hill, New York (1970).
10. R. W. Berry, D. M. Hall, and M. T. Harris, eds., *Thin Film Technology*, Van Nostrand, Princeton (1968).
11. J. L. Vossen and W. Kern, eds., *Thin Film Processes*, Academic Press, New York (1978).
12. J. C. Anderson, ed., *The Use of Thin Films in Physical Investigations*, Academic Press, New York (1966).
13. Inderjeet Kaur, D. K. Pandya, and K. L. Chopra, *J. Electrochem Soc.* **127**, 943 (1980).
14. K. L. Chopra, M. R. Randlett, and R. H. Duff, *Philos. Mag.* **16**, 261 (1967).
15. L. Esaki and C. L. Chang, *Thin Solid Films* **36**, 285 (1976).
16. D. R. Arthur, *J. Vac. Sci. Technol.* **16**, 273 (1979).
17. C. F. Powell, in *Vapor Deposition*, C. F. Powell, J. H. Oxley, and J. M. Blocher, Jr., eds., Wiley, New York (1966), p. 249.
18. W. M. Feist, S. R. Steele, and D. W. Readey, in *Physics of Thin Films*, Vol. 5, G. Hass and R. E. Thun, eds., Academic Press, New York (1969), p. 237.
19. T. L. Chu and K. N. Singh, *Solid State Electron* **19**, 837 (1976).
20. N. C. Sharma, D. K. Pandya, H. K. Sehgal, and K. L. Chopra, *Thin Solid Films* **59**, 157 (1979).
21. I. Kaur, Growth of Compound Semiconductor Films from Aqueous Solutions, Ph.D. Thesis, Indian Institute of Technology, Delhi (1981).
22. R. C. Kainthla, Solution Growth of Semiconducting Alloy Films: Their Study and Application in Photoelectrochemical Cells, Ph.D. Thesis, Indian Institute of Technology, Delhi (1980).

23. N. C. Sharma, Solution-Grown Variable-Band-Gap $Pb_{1-x}Hg_xS$ Films, Ph.D. Thesis, Indian Institute of Technology, Delhi (1978).
24. N. C. Sharma, R. C. Kainthla, D. K. Pandya, and K. L. Chopra, *Thin Solid Films* **60**, 55 (1979).
25. R. C. Kainthla, D. K. Pandya, and K. L. Chopra, *J. Electrochem. Soc.* **129**, 99 (1982).
26. K. L. Chopra, R. C. Kainthla, D. K. Pandya, and A. P. Thakoor, in *Physics of Thin Films*, Vol. 12, M. H. Francombe, ed., Academic Press, New York (1982).
27. I. Kaur and K. L. Chopra, Paper presented at the 5th International Thin Film Congress, Herzlia, Israel, September 1981.
28. J. R. Szedon, F. A. Shirlard, W. J. Biter, T. W. O'Keeffe, J. A. Stoll, and S. J. Fonash, Cds/Cu$_2$S Heterojunction Solar Cells Research, Final Report No. 78-9F3-CADSO-R6, Westinghouse R and D Center (1979).
29. F. A. Lowenheim, ed., *Modern Electroplating*, Wiley, New York (1974).
30. F. A. Lowenheim, ed., *Electroplating*, McGraw-Hill, New York (1978).
31. A. S. Baranski and W. R. Fawcett, *J. Electrochem. Soc.* **127**, 766 (1980).
32. L. Young, *Anodic Oxide Films*, Academic Press, New York (1961).
33. M. A. Coler, in *Encyclopedia of Chemical Technology (Kirk-Othmer)*, Vol. 8, H. F. Mark, J. J. McKetta, Jr., and D. F. Othmer, eds., Wiley, New York (1965), p. 23.
34. N. Nakayama, *Jpn. J. Appl. Phys.* **8**, 450 (1969).
35. S. Saxena, D. K. Pandya, and K. L. Chopra, *Thin Solid Films* **94**, 223 (1982).
36. A. N. Broers, Inst. Phys. Conf. Ser. No. 40, p. 155 (1978); A. N. Broers and M. Hatzakis, *Sci. Am.* **220**, 34 (November 1972).
37. H. I. Smith and D. C. Flanders, *J. Vac. Sci. Technol.* **17**, 533 (1980).
38. M. J. Bowden, *CRC Crit. Rev. Solid State Sci.* p. 223 (February 1979).
39. R. F. Pease, *Comtemp. Phys.* **22**, 265 (1981).

References for Chapter 2

1. A. Herpin, *Compt. Rend.* **225**, 182 (1947).
2. W. Weinstein, *Vacuum* **4**, 3 (1954).
3. H. A. Macleod, *Thin Film Optical Filters*, Adam Hilger Ltd., London (1969).
4. J. T. Cox and G. Hass, in *Physics of Thin Films*, Vol. 2, G. Hass and R. E. Thun, eds., Academic Press, New York (1964), p. 239.
5. A. Mussett and A. Thelen, *Prog. Opt.* **8**, 203 (1970).
6. R. B. Muchmore, *J. Opt. Soc. Am.* **38**, 20 (1948).
7. V. N. Yadava, Optical Behavior of Thin Films of Variable Optical Constants, Ph.D. Thesis, Indian Institute of Technology, Delhi, India (1973).
8. R. Jacobsson and J. D. Martensson, *Appl. Opt.* **5**, 29 (1966).
9. G. Hass, *J. Opt. Soc. Am.* **45**, 945 (1955).
10. S. Penselin and A. Stendel, *Z. Phys.* **142**, 21 (1955).
11. L. I. Epstein, *J. Opt. Soc. Am.* **42**, 806 (1952).
12. P. Kard, E. A. Nesmelov, and G. P. Konyukhov, *Eesti NSV Tead. Akad. Toim., Fuus.-Mat.* **17**, 314 (1968); E. A. Nesmelov and G. P. Konyukhov, *Opt. Spektrosk.* **31**, 123 (1968).
13. P. W. Baumeister and F. A. Jenkins, *J. Opt. Soc. Am.* **47**, 57 (1957).
14. R. Jacobson, *J. Opt. Soc. Am.* **54**, 422 (1964).
15. L. Young, *Appl. Opt.* **6**, 297 (1967).
16. A. Thelen, *J. Opt. Soc. Am.* **53**, 1266 (1963).
17. T. K. Kumari, Interference Filters, Masters Thesis, Indian Institute of Technology, Delhi, India (1980).

18. S. D. Smith, *J. Opt. Soc. Am.* **48**, 43 (1958).
19. A. F. Turner *et al.*, *Infrared Transmission Filters*, Bausch and Lomb Tech. Rep., Nos. 1–6 (1953).
20. A. F. Turner, *J. Opt. Soc. Am.* **42**, 878 (A), 1952.
21. P. H. Berning and A. F. Turner, *J. Opt. Soc. Am.* **47**, 230 (1957).
22. A. F. Turner, *J. Phys. Radium* **11**, 444 (1950).
23. B. H. Billings, *J. Opt. Soc. Am.* **40**, 471 (1950).
24. F. Abeles, *J. Phys. Radium* **11**, 310, 453 (1950).
25. P. B. Clapham, M. J. Downs, and R. J. King, *Appl. Opt.* **8**, 1965 (1969).
26. M. Banning, *J. Opt. Soc. Am.* **37**, 688, 792 (1947).
27. P. K. Tien, *Sci. Am.* **230**, 28 (April 1974).
28. P. K. Tien, *Appl. Opt.* **10**, 2395 (1971).
29. T. Tamir, ed., *Topics in Applied Physics*, Vol. 7: *Integrated Optics*, Springer-Verlag, Berlin (1975).
30. R. Schubert and J. H. Harris, *J. Opt. Soc. Am.* **61**, 154 (1971).
31. D. C. Flanders, H. Kogelnik, R. V. Schmidt, and C. V. Shank, *Appl. Phys. Lett.* **24**, 194 (1974).
32. P. K. Tien, R. Ulrich, and R. J. Martin, *Appl. Phys. Lett.* **14**, 291 (1969).
33. M. L. Dakss, L. Kuhn, P. F. Heidrich, and B. A. Scott, *Appl. Phys. Lett.* **16**, 523 (1970).
34. P. K. Tien and R. J. Martin, *Appl. Phys. Lett.* **18**, 398 (1971).
35. J. M. Hammer and W. Phillips, *Appl. Phys. Lett.* **24**, 545 (1974).
36. L. Kuhn, M. L. Dakss, P. L. Heidrich, and B. A. Scott, *Appl. Phys. Lett.* **17**, 265 (1970).
37. H. F. Taylor, *J. Appl. Phys.* **44**, 3257 (1973).
38. P. K. Tien, R. J. Martin, R. Wolfe, R. C. LeCraw, and S. L. Blank, *Appl. Phys. Lett.* **21**, 397 (1972).
39. F. K. Reinhart, *J. Appl. Phys.* **39**, 3426 (1968); F. K. Reinhart, D. F. Nelson, and J. McKenna, *Phys. Rev.* **177**, 1208 (1969).
40. L. A. D'Asaro, in *Physics and Technology of Semiconductor Light Emitters and Detectors*, A. Prova, ed., North-Holland, Amsterdam (1973); E. Garmire, in *Topics in Applied Physics*, Vol. 7: *Integrated Optics*, T. Tamir, ed., Springer-Verlag, Berlin (1975).
41. F. A. Blum, K. L. Lawley, W. C. Scott, and W. C. Holton, *Appl. Phys. Lett.* **24**, 430 (1974).
42. G. E. Stillman, C. M. Wolfe, A. K. Foyt, and W. T. Lindley, *Appl. Phys. Lett.* **24**, 8 (1974).
43. S. Somekh, E. Garmire, A. Yariv, H. Garvin, and R. Hunsperger, *Appl. Opt.* **12**, 455 (1973).
44. H. Stoll, A. Yariv, R. Hunsperger, and G. Tangonan, *Appl. Phys. Lett.* **23**, 664 (1973).
45. O. S. Heavens, *Optical Properties of Thin Solid Films*, Butterworths, London (1955).

References for Chapter 3

1. R. H. Bube, *Photoconductivity of Solids*, Wiley, New York (1967).
2. N. C. Sharma, D. K. Pandya, A. K. Mukherjee, H. K. Sehgal, and K. L. Chopra, *Appl. Opt.* **16**, 2945 (1977).
3. I. Kaur, D. K. Pandya, and K. L. Chopra, *J. Electrochem. Soc.* **127**, 943 (1980).
4. R. C. Kainthla, D. K. Pandya, and K. L. Chopra, *J. Electrochem. Soc.* **129**, 99 (1982).
5. L. Ley and M. Cardona, eds., *Topics in Applied Physics*, Vol. 27: *Photoemission in Solids*, Springer-Verlag, Berlin (1979).
6. D. Kossel, K. Deutscher, and K. Hirschberg, in *Advances in Electronics and Electron Physics*, Vol. 28A, L. Marton, ed., Academic Press, London (1969), p. 419; *Photoelectronic Image Devices*: Proc. of the 4th Symposium held at Imperial College, London, Sept. 16–20, 1968, J. D. McGee, D. McMullan, E. Kahan, and B. L. Morgan, eds.

7. D. Kossel, K. Deutscher, and K. Hirschberg in *Physics of Thin Films*, Vol. 5, G. Hass and R. F. Thun, eds. Academic Press, New York (1969).

8. S. Weber, Telefunken Report on Reflective Interference Cathodes (1968).

9. C. H. A. Syms, in *Advances in Electronics and Electron Physics*, Vol. 28A, L. Marton ed., Academic Press, London (1969); p. 399; *Photoelectronic Image Devices*: Proc. of the 4th Symposium held at Imperial College, London, Sept. 16–20, 1968, J. D. McGee, D. McMullan, E. Kahan, and B. L. Morgan, eds.

10. M. B. Allenson, P. G. R. King, M. C. Rowland, G. J. Steward, and C. H. A. Syms, *J. Phys. D: Appl. Phys.* **5**, L89 (1972).

11. R. U. Martinelli and D. G. Fischer, *Proc. IEEE* **62**, 1339 (1974).

12. A. G. Milnes and D. L. Feucht, *Heterojunctions and Metal–Semiconductor Junctions*, Academic Press, New York (1972).

13. B. L. Sharma and R. K. Purohit, *Semiconductor Heterojunctions*, Pergamon Press, Oxford (1974).

14. S. M. Sze, *Physics of Semiconductor Devices*, Wiley, New York (1969).

15. R. Hill, in *Active and Passive Thin Film Devices*, T. J. Coutts, ed., Academic Press, London (1978), p. 491.

16. R. Hill, in *Active and Passive Thin Film Devices*, T. J. Coutts, ed., Academic Press, London (1978), p. 496.

17. P. T. Landsberg, *Solid State Electron.* **18**, 1043 (1976).

18. J. J. Loferski, Proc. of the 12th Photovoltaic Specialists Conference, Baton Rouge (1976), p. 957.

19. J. A. Cape, J. S. Harris, Jr., and R. Sahai, Proc. of the 13th IEEE Photovoltaic Specialists Conference, Washington, D. C. (1978), p. 881.

20. G. W. Masden and C. E. Backus, Proc. of the 13th IEEE Photovoltaic Specialists Conference, Washington, D.C. (1978), p. 853.

21. M. P. Vecchi, *Solar Energy* **22**, 383 (1979).

22. M. F. Lamorte and D. Abbott, *Solid-State Electron.* **22**, 467 (1979).

23. H. Gerischer, in *Topics in Applied Physics*, Vol. 31: *Solar Energy Conversion*, B. O. Seraphin, ed., Springer-Verlag, Berlin (1979), p. 115.

24. A. B. Ellis, S. W. Kaiser, and M. S. Wrighton, *J. Am. Chem. Soc.* **98**, 6855 (1976).

25. H. Gerischer, *Electroanal. Chem. Interfacial Electrochem.* **58**, 263 (1975).

26. S. M. Bedair, M. F. Lamorte, and J. R. Hanser, *Appl. Phys. Lett.* **34**, 38 (1979).

27. L. M. Frass and R. C. Knechtli, Proc. of the 13th IEEE Photovoltaic Specialists Conference, Washington, D. C. (1978), p. 885.

28. M. Ariezo and J. J. Loferski, *Proc. of the 13th IEEE Photovoltaic Specialists Conference*, Washington, D.C. (1978), p. 898.

29. K. L. Chopra and S. R. Das, *Thin Film Solar Cells*, Plenum Press, New York (1982).

30. B. Kazan and M. Knoll, *Electronic Image Storage*, Academic Press, New York (1968).

31. L. M. Biberman and S. Nudelman, eds., *Photoelectronic Imaging Devices*, Plenum Press, New York (1971).

32. Y. Kajiyama, T. Kawahara, and T. Hirayama, *NEC Res. Dev.* No. 11, p. 133 (1968).

33. A. H. Boerio, R. R. Beyer, and G. W. Goetze, in *Advances in Electronics and Electron Physics*, Vol. 22A, L. Marton, ed., Academic Press, London (1966), p. 229; *Photoelectronic Image Devices*: Proc. of the 3rd Symposium held at Imperial College, London, Sept. 20–24, 1965, J. D. McGee, D. McMullan, and E. Kahan, eds.

34. J. H. Dessauer and H. E. Clark, eds., *Xerophotography and Related Processes*, Focal Press, London (1965); R. M. Schaffert. *Electrophotography*, Focal Press, London (1965).

35. S. van Houten, in *Solid State Devices*, Proc. of the 3rd European Solid State Device Research Conf. (ESSDERC) held in Munich, September 18–21, 1973, p. 131, The Institute of Physics, London (1974).

36. P. Goldberg, ed., *Luminescence of Inorganic Solids*, Academic Press, New York (1966).

37. G. Destriau, *J. Chem. Phys.* **33**, 620 (1936); *Philos. Mag.* **38**, 700 (1947).
38. O. W. Lossev, *Wireless World Radio Rev.* **15**, 93 (1924); *Compt. Rend. Acad. Sci. USSR* **29**, 363 (1940).
39. T. Inoguchi, M. Takeda, Y. Kakihara, Y. Nakata, and M. Yoshida, *SID Int. Symp. Dig.* p. 84 (1974).
40. F. H. Nicoll, *Photoelectronic Materials and Devices*, Van Nostrand, New Jersey (1965), p. 313.
41. M. Takeda, Y. Kakihara, M. Yoshida, M. Kawaguchi, H. Kishishita, Y. Yamauchi, T. Inoguchi, and S. Mito, *SID Int. Symp. Dig.*, Section 7.8 (1975).
42. A. G. Fischer and H. I. Moss, *J. Appl. Phys.* **34**, 2112 (1963).
43. A. H. Rosenthal, *Proc. IRE* **28**, 203 (1940).
44. P. G. R. King, *Proc. IEE (London)* **93A**, 171 (1946).
45. S. K. Deb, *Appl. Opt. Suppl.* **3**, 192 (1969).
46. S. K. Deb and R. F. Shaw, US Patent No. 3,521,941 (1970).
47. J. J. Robillard, US Patent No. 3,457,069 (1969).
48. E. M. Kosower and E. J. Poziomek, *J. Am. Chem. Soc.* **86**, 5515 (1964); E. M. Kosower and J. L. Cotter, *J. Am. Chem. Soc.* **86**, 5524 (1964).
49. W. H. Hansen, R. A. Osteryoung, and T. Kuwana, *J. Am. Chem. Soc.* **88**, 1062 (1966).
50. I. F. Chang, B. L. Gilbert, and T. I. Sun, IBM Report RC 5179, Dec. 1974; *J. Electrochem. Soc.* **122**, 955 (1975).
51. I. F. Chang and B. L. Gilbert, *IBM Tech. Discl. Bull.* **17**, 1050 (1974).
52. W. J. Horkans and L. T. Romanikiw, *IBM Tech. Discl. Bull.* **17**, 2517 (1975).
53. J. A. Hall and J. J. McCann, US Patent No. 3,692,388 (1972).
54. G. D. Jones and R. E. Friedrich, US Patent No. 3,283,656 (1966).
55. F. H. Smith, British Patent No. 328,017 (1929).
56. H. deKoster, US Patent No. 362,410 (1971).
57. D. J. Berets, US Patent No. 3,879,108 (1975).
58. H. Witzke and S. E. Schualterly, US Patent No. 3,840,287 (1974).
59. E. Sauser and S. A. Ebauches, US Patent No. 3,836,229 (1974).
60. E. E. Schualterly, US Patent No. 3,840,288 (1974).
61. P. Talmey, US Patent No. 2,319,765 (1943).
62. G. A. Castellion, US Patent No. 3,807,832 (1974).
63. D. Chen and J. D. Zook, *Proc. IEEE* **63**, 1207 (1975).
64. D. Maydan, *Bell Syst. Tech. J.* **50**, 1761 (1971); J. Corcoran and H. Ferris, SPIE 21st International Tech. Symp. and Instrument Display, San Diego, California (1977).
65. K. Bulthuis, M. G. Carasso, J. P. J. Heemskerk, P. J. Kivits, W. J. Kleuters, and P. Zalm, *IEEE Spectrum* **16**, 26 (August 1979).
66. G. C. Kenny, D. Y. K. Lou, R. McFarlane, A. Y. Chan, J. S. Nadan, T. R. Kohler, J. G. Wagner, and F. Zernike, *IEEE Spectrum* **16**, 33 (February 1979).
67. M. Terao, K. Shigematsu, M. Ojima, Y. Taniguchi, S. Horigome, and S. Yonezawa, Proc. of the 11th Conf. on Solid State Devices, Tokyo (1979); *Jpn. J. Appl. Phys.* **19**, Supplement 19-1, 579 (1980).
68. J. A. Rajchman, *J. Appl. Phys.* **41**, 1376 (1970).
69. J. A. Rajchman, *Appl. Opt.* **9**, 2269 (1970).
70. W. C. Stewart and L. S. Cosentino, *Appl. Opt.* **9**, 2271 (1970).
71. T. D. Beard, W. P. Bleha, and S. Y. Wong, *Appl. Phys. Lett.* **22**, 90 (1973).
72. H. J. Williams, R. C. Sherwood, E. G. Foster, and F. M. Kelley, *J. Appl. Phys.* **28**, 1181 (1957).
73. T. C. Lee and D. Chen, *Appl. Phys. Lett.* **19**, 62 (1971).
74. D. Chen, J. F. Ready, and E. Bernal, *J. Appl. Phys.* **39**, 3916 (1968).
75. H. Haskal, G. E. Bernal, and D. Chen, *Appl. Opt.* **13**, 866 (1974).
76. J. J. Anodei, and R. S. Mezrich, *Appl. Phys. Lett.* **15**, 45 (1969).

77. G. Decker, H. Herold, and H. Röhr, *Appl. Phys. Lett.* **20**, 490 (1972).
78. J. C. Urbach, in *Topics in Applied Physics*, Vol. 20: *Holographic Recording Materials*, H. M. Smith, ed., Springer-Verlag, Berlin (1977), p. 161.
79. B. Singh, S. Rajagopalan, P. K. Bhat, D. K. Pandya, and K. L. Chopra, *Solid State Commun.* **29**, 167 (1979).
80. B. Singh, S. Rajagopalan, P. K. Bhat, D. K. Pandya, and K. L. Chopra, *J. Non-Cryst. Solids* **35**, 1053 (1980).
81. K. L. Chopra, K. S. Harshvardhan, S. Rajagopalan, and L. K. Malhotra, *Solid State Commun.* **40**, 387 (1981).
82. K. L. Chopra, K. S. Harshvardhan, S. Rajagopalan, and L. K. Malhotra, *Appl. Phys. Lett.* **40**, 428 (1982).
83. K. L. Chopra, L. K. Malhotra, K. S. Harshvardhan, and B. Singh, paper presented at the ECS Meeting, Montreal, Canada, May 1982.
84. J. Feinleib, J. Denneufville, S. C. Moss, and S. R. Ovshinsky, *Appl. Phys. Lett.* **18**, 254 (1971).
85. S. A. Keneman, *Appl. Phys. Lett.* **19**, 205 (1971).
86. Y. Ohmachi and T. Igo, *Appl. Phys. Lett.* **20**, 506 (1972).
87. V. I. Mandrosov, E. I. Pik, and G. A. Sobolev, *Opt. Spectrosc.* **34**, 695 (1973); **35**, 75 (1973).
88. S. A. Keneman, *Thin Solid Films* **21**, 281 (1974).
89. M. I. Kostyshin, E. P. Krasnojonov, V. A. Makeev, and G. A. Sovolev, in *Applications of Holography*, Proc. of the International Symp. of Holography, J. C. Vienot, J. Bulabois, and J. Pasteur, eds., Univ. Besancon, France (1970), paper No. 11.7.
90. S. A. Keneman, G. W. Taylor, A. Miller, and W. H. Fonger, *Appl. Phys. Lett.* **17**, 173 (1970).
91. W. D. Smith and C. E. Land, *Appl. Phys. Lett.* **20**, 169 (1972).
92. H. Kiemle and U. Wolff, Annual Meeting of the Optical Society of America, Sept. 28–Oct. 2, 1970.
93. J. D. Margerum, J. Nimoy, and S. Y. Wong, *Appl. Phys. Lett.* **17**, 51 (1970).
94. G. Assouline, M. Hareng, and E. Leiba, *Proc. IEEE* **59**, 1355 (1971).
95. S. G. Lipson and P. Nisenson, *Appl. Opt.* **13**, 2052 (1974).
96. See for example: D. E. Carlson, *J. Vac. Sci. Tech.* **20**, 290 (1982); W. E. Spear, P. G. LeComber, A. J. Snell, and R. A. Gibson, *Thin Solid Films* **90**, 359 (1982); and H. Okamoto, Y. Nitta, T. Adachi, and Y. Hamakawa, *Solar Energy Materials* **2**, 313 (1980) and references cited therein.

References for Chapter 4

1. K. L. Chopra, *Thin Film Phenomena*, McGraw-Hill, New York (1969).
2. L. I. Maissel and R. Glang, eds., *Handbook of Thin Film Technology*, McGraw-Hill, New York (1970).
3. T. J. Coutts, ed., *Active and Passive Thin Film Devices*, Academic Press, London (1978).
4. L. Holland, ed., *Thin Film Microelectronics: The Preparation and Properties of Components and Circuit Arrays*, Chapman and Hall, London (1965).
5. N. Schwartz and R. W. Berry, in *Physics of Thin Films*, Vol. 2, G. Hass and R. E. Thun, eds., Academic Press, New York (1964), p. 363.
6. K. Fuchs, *Proc. Cambridge Philos. Soc.* **34**, 100 (1938).
7. F. Savornin, *C. R. Acad. Sci.* **248**, 2458 (1959).
8. A. F. Mayadas and M. Shatzkes, *Phys. Rev. B* **1**, 1382 (1970).
9. C. A. Neugebauer, in *Physics of Thin Films*, Vol. 2, G. Hass and R. E. Thun, eds., Academic Press, New York (1964), p. 1.

10. C. A. Neugebauer and R. H. Wilson, in *Basic Problems in Thin Film Physics*, R. Niedermayer and H. Mayer, eds., Vandenhoeck and Ruprecht, Göttingen (1966), p. 579.

11. W. Kohn, *Phys. Rev.* **110**, 857 (1958).

12. K. L. Chopra, *J. Appl. Phys.* **36**, 655 (1965).

13. S. Pakswer and K. Pratinidhi, *J. Appl. Phys.* **34**, 711 (1963).

14. K. L. Chopra, M. R. Randlett, and R. H. Duff, *Philos. Mag.* **16**, 261 (1967).

15. H. J. Schuetze, H. W. Ehlbeck, and G. G. Doerbeck, *Transactions of the 10th National Vacuum Symposium*, Pergamon Press, New York (1963), p. 434.

16. D. Gerstenberg and C. J. Calbick, *J. Appl. Phys.* **35**, 402 (1964).

17. P. M. Schaible and L. I. Maissel, *Transactions of the 9th National Vacuum Symposium*, Macmillan, New York (1962), p. 190.

18. F. J. Hemmer, C. Feldman, and W. T. Layton, *Proc. Natl. Electron, Conf.* **20**, 201 (1964).

19. C. M. Jackson, J. G. Dunleavy, and A. M. Hall, *Proceedings of the Electronic Components Conference*, 1961, p. 36-1.

20. W. Himes, B. F. Stout, and R. E. Thun, *Transactions of the 9th National Vacuum Symposium*, Macmillan, New York (1962), p. 144.

21. T. K. Lakshmanan, *Transactions of the 8th National Vacuum Symposium*, Pergamon Press, New York (1961), p. 868.

22. I. H. Pratt, *Proc. Natl. Electronic Conf.* **20**, 215 (1964).

23. M. A. Foley, *Proceedings of the Electron Components Conference*, 1967, p. 422.

24. M. Beckerman and R. E. Thun, *Transactions of the 8th National Vacuum Symposium*, Pergamon Press, New York (1961), p. 905.

25. N. C. Miller and G. A. Shirn, *Appl. Phys. Lett.* **10**, 86 (1967).

26. W. O. Freitag and V. R. Weiss, *Res. Dev.* p. 44, (August 1967).

27. W. Himes, C. M. Stout, and R. E. Thun, *Transactions of the 9th National Vacuum Symposium*, Macmillan, New York (1962), p. 144; R. Glang, R. A. Holmwood, and S. R. Herd, *J. Vac. Sci. Technol.* **4**, 163 (1967).

28. R. O. Grisdale, A. C. Pfister, and G. K. Teal, *Bell System Tech. J.* **30**, 271 (1951).

29. R. E. Aitchison, *Aust. J. Appl. Sci.* **5**, 10 (1954); D. F. A. MacLachlam, L. S. Phillips, K. R. Honick, and G. V. Planer, *J. Brit. Inst. Radio Eng.* **21**, 221 (1961).

30. H. W. Burkett, *J. Br. Inst. Radio Eng.* **21**, 301 (1961); S. A. Halaby, L. V. Gregor, and S. M. Rubens, *Electro. Tech.* *(N.Y.)* **72(3)**, 95 (1963).

31. F. S. Maddocks and R. E. Thun, *J. Electrochem. Soc.* **109**, 99 (1962).

32. H. J. Degenhart and I. H. Pratt, *Transactions of the 8th National Vacuum Symposium*, Pergamon Press, New York (1961), p. 859.

33. F. G. Peters, *Am. Ceram. Soc. Bull.* **45**, 1017 (1966).

34. D. A. McLean, N. Schwartz, and E. D. Tidd, *Proc. IEEE* **52**, 1450 (1964).

35. D. A. McLean and F. E. Rosztoczy, *Electrochem. Technol.* **4**, 523 (1966).

36. R. M. Valleta and W. A. Pliskin, *J. Electrochem. Soc.* **114**, 944 (1966).

37. H. N. Keller, C. T. Kemmerer and C. L. Naegele, *IEEE Trans. Parts Mater. Packag.* **3**, 97 (1967).

38. I. Blech, H. Sello, and L. V. Gregor, in *Handbook of Thin Film Technology*, L. I. Maissel and R. Glang, eds., McGraw-Hill, New York (1970), p. 23-1.

39. J. K. Howard and R. F. Ross, *Appl. Phys. Lett.* **11**, 85 (1967).

40. F. M. d'Heurle and R. Rosenburg, in *Physics of Thin Films*, Vol. 7, G. Hass, M. H. Francombe, and R. W. Hoffman, eds., Academic Press, New York (1973), p. 257.

41. P. B. Ghate, *Appl. Phys. Lett.* **11**, 14 (1967).

42. J. A. Cunningham, *Solid State Electron.* **8**, 735 (1965).

43. M. P. Lepselter, *Bell System Tech. J.* **45**, 233 (1966).

44. A. C. Tickle, *Thin Film Transistors: A New Approach to Microelectronics*, Wiley, New York (1969).

45. P. K. Weimer, in *Field Effect Transistors*, J. T. Wallmark and H. Johnson, eds., Prentice-Hall, Englewood Cliffs, New Jersey (1966); in *Physics of Thin Films*, Vol. 2, G. Hass and R. E. Thun, eds., Academic Press, New York (1964), p. 147.

46. P. K. Weimer, *Proc. IRE* **50**, 1462 (1962).

47. F. V. Shallcross, *Proc. IEEE* **51**, 851 (1963).

48. P. K. Weimer, *Proc. IEEE* **52**, 608 (1964).

49. V. L. Frantz, *Proc. IEEE* **53**, 760 (1965).

50. H. A. Klasens and H. Koelmans, *Solid-State Electron.* **7**, 701 (1964).

51. W. B. Pennebaker, *Solid-State Electron.* **8**, 509 (1965).

52. J. F. Skalski, *Proc. IEEE* **53**, 1792 (1965).

53. T. P. Brody and H. E. Kunig, *Appl. Phys. Lett.* **9**, 259 (1966).

54. H. Borkan and P. K. Weimer, *RCA Rev.* **24**, 153 (1963).

55. H. L. Wilson and W. Gutierrez, *Proc. IEEE* **55**, 415 (1967).

56. A. Waxman, *Electronics* **41**, 88 (March 1968).

57. C. M. Mueller and P. H. Robinson, *Proc. IEEE* **52**, 1487 (1964).

58. R. S. Muller and R. Zuleeg, *J. Appl. Phys.* **35**, 1550 (1964).

59. F. Huber, *Solid-State Electron.* **5**, 410 (1962).

60. J. C. Anderson, *Thin Solid Films* **12**, 1 (1972).

61. H. Basseches and D. Gerstenburg, *Thin Solid Films* **12**, 295 (1972).

62. D. A. McLean, *Thin Solid Films* **8**, 1 (1971).

63. D. A. McLean, N. Schwartz, and E. D. Tidd, *Proc. IEEE* **52**, 1450 (1964).

64. L. V. Gregor, *Proc. IEEE* **59**, 1390 (1971); P. L. Kirby, *Thin Solid Films* **50**, 211 (1978).

65. P. K. Weimer, G. Sadasiv, L. Merray-Horvath, and W. S. Homa, *Proc. IEEE* **54**, 354 (1966).

66. P. K. Weimer, G. Sadasiv, J. E. Meyer, Jr., L. Merray-Horvath, and W. S. Pike, *Proc. IEEE* **55**, 1591 (1967).

67. Fang-Chen Luo and A. William, *IEEE Trans. Electron Devices* **ED-27**, 223 (1980).

68. T. P. Brody, *Proc. SID* **17**, 39 (1st Quarter 1976); T. P. Brody, F. C. Luo, Z. P. Szepesi, and D. H. Davies, *IEEE Trans. Electron Devices* **ED-22**, 739 (1975).

69. J. C. Erskine and P. A. Snopko, *IEEE Trans. Electron Devices* **ED-26**, 802 (1979).

70. K. O. Fugate, *IEEE Trans. Electron Devices* **ED-24**, 909 (1977).

71. K. C. Gupta and Amarjit Singh, *Microwave Integrated Circuits*, Wiley, New York (1974).

72. C. E. Jowett, *The Engineering of Microelectronic Thin and Thick Films*, Macmillan (1976), p. 63; M. Caulton, *Proc. IEEE* **59**, 1481 (1971).

73. M. Caulton, J. J. Hughes, and H. Sobol, *RCA Rev.* **27**, 377 (1966).

74. E. G. Lean and A. N. Broers, *Microwave J.* **13**, 17 (March 1970).

75. K. Wasa and S. Hayakawa, in *Vacuum–Surfaces–Thin Films*, K. L. Chopra and T. C. Goel, eds., Vanity Books, Dehli (1981), p. 95.

76. P. Hartemann and E. Dieulesaint, *Electron. Lett.* **5**, 657 (1969).

77. P. Hartemann and E. Dieulesaint, *Electron. Lett.* **5**, 219 (1969).

78. I. Kaufman and J. W. Foltz, *Proc. IEEE (Lett.)* **57**, 2081 (1969).

79. V. O. Blackledge and I. Kaufman, *Appl. Phys. Lett.* **15**, 127 (1969).

80. G. S. Kino and H. Matthews, *IEEE Spectrum* **8**, 22 (August 1971).

81. R. M. White, *Proc. IEEE* **58**, 1238 (1970).

82. E. G. S. Paige, in *Solid State Devices*, Proceedings of the 2nd European Solid State Device Research Conference (ESSDERC) held at the University of Lancaster, September 12–15, 1972, p. 39, The Institute of Physics, London (1973).

83. G. S. Hobson, *Charge Transfer Devices*, Edward Arnold, London (1978).

84. F. L. J. Sangster and K. Teer, *IEEE J. Solid State Circuits* **SC-4**, 131 (1969).

85. W. S. Boyle and G. E. Smith, *Bell System Tech. J. Briefs* **49**, 587 (1970).

86. J. E. Carnes, in *Solid State Devices*, Proceedings of the 3rd European Solid State Device Research Conference (ESSDERC) held in Munich, September 18–21, 1973, p. 83, The Institute of Physics, London (1974).

87. G. F. Amelio, *Sci. Am.* **230**, 22 (February 1974).
88. G. R. Witt, *Thin Solid Films* **22**, 133 (1974).
89. R. L. Parker and A. Krinsky, *J. Appl. Phys.* **34**, 2700 (1963).
90. Z. H. Meiksin and R. A. Hudzinski, *J. Appl. Phys.* **38**, 4490 (1967).
91. S. C. Chang, *IEEE Trans. Electron Devices* **ED-26**, 1875 (1979).
92. G. N. Advani and A. G. Gordan, *J. Electron. Mater.* **9**, 29 (1980).
93. G. N. Advani, Y. Komem, J. Hasenkopf, and A. G. Gordan, *Sensors Actuators* **2**, 139 (1981).
94. I. Lundström, *Sensors Actuators* **1**, 403 (1981).
95. T. L. Poteat and B. Lalevic, *IEEE Trans. Electron Devices* **ED-29**, 123 (1982).

References for Chapter 5

1. R. F. Soohoo, *Magnetic Thin Films*, Harper and Row, New York (1965).
2. M. Prutton, *Thin Ferromagnetic Films*, Butterworths, London (1964).
3. K. L. Chopra, *Thin Film Phenomena*, McGraw-Hill, NewYork (1969).
4. E. W. Pugh and T. O. Mohr, in *Thin Films*, Chapman and Hall (1964).
5. M. S. Cohen, in *Handbook of Thin Film Technology*, L. I. Maissel and R. Glang, eds., McGraw-Hill, New York (1970), Section 17-1.
6. L. Neel, *Compt. Rend.* **237**, 1468 (1953).
7. G. Robinson, *J. Phys. Soc. Jpn.* **17** (Suppl. B-I), 588 (1962).
8. R. J. Prosen, Y. Gondo, and B. E. Gran, *J. Appl. Phys.* **35**, 826 (1964).
9. S. Middelhoek, *J. Appl. Phys.* **34**, 1054 (1963).
10. R. J. Spain and H. W. Fuller, *J. Appl. Phys.* **37**, 953 (1966); N. Saito, H. Fujiwara, and Y. Sugita, *J. Phys. Soc. Jpn.* **19**, 421, 1116 (1964).
11. E. C. Stoner and E. P. Wohlfarth, *Philos. Trans. Roy. Soc. London Ser. A.* **240**, 599 (1948).
12. E. M. Bradley and M. Prutton, *J. Electron. Control* **6**, 81 (1959).
13. M. Prutton, *Br. J. Appl. Phys.* **11**, 335 (1960).
14. D. O. Smith and K. J. Harte, *J. Appl. Phys.* **33**, 1399 (1962).
15. H. W. Fuller and M. E. Hale, *J. Appl. Phys.* **31**, 238 (1960).
16. J. E. Schwenker and T. R. Long, *J. Appl. Phys.* **33**, 1099 (1962).
17. A. V. Pohm and E. N. Mitchell, *IRE Trans. Electron. Comput.* **EC-9**, 308 (1960).
18. R. V. Telesnin, E. T. Elicheva, O. S. Kolotov, T. N. Nikitina, and V. A. Pogozhev, *Phys. Status Solidi* **14**, 371 (1966).
19. S. Middelhoek and D. Wild, *IBM J. Res. Dev.* **11**, 93 (1967).
20. C. D. Olson and A. V. Pohm, *J. Appl. Phys.* **29**, 274 (1958).
21. P. D. Barker and E. J. Torok, *J. Appl. Phys.* **37**, 1363 (1966).
22. K. U. Stein and E. Feldtkeller, *IEEE Trans. Magn.* **MAG-2**, 184 (1966).
23. I. Danylchuk, A. J. Perneski, and M. W. Sagal, *Proceedings of the Intermagnetic Conference*, New York (1964), p. 5-4-1, IEEE Press, New York (1964).
24. G. Y. Fan and J. H. Greiner, *J. Appl. Phys.* **39**, 1216 (1968).
25. W. K. Unger, *Int. J. Magn.* **3**, 43 (1972).
26. R. C. Sherwood, E. A. Nesbitt, J. H. Wernick, D. D. Bacon, A. J. Kurtzig, and R. Wolfe, *J. Appl. Phys.* **42**, 1704 (1971).
27. G. B. Street, E. Sawatzky, and K. Lee, *J. Appl. Phys.* **44**, 410 (1973).
28. H. Weider, S. I. Lavenberg, G. J. Fan, and R. A. Burn, *J. Appl. Phys.* **42**, 3458 (1971).
29. H. Wieder and R. A. Burn, *IEEE Trans. Magn.* **MAG-9**, 471 (1973).
30. H. Wieder and R. A. Burn, *J. Appl. Phys.* **44**, 1774 (1973).
31. N. Goldberg, *IEEE Trans. Magn.* **MAG-3**, 605 (1967).
32. P. Chaudhari, J. J. Cuomo, and R. J. Gambino, *Appl. Phys. Lett.* **22**, 337 (1973).
33. H. Rubinstein, R. M. Hornreich, J. Teixeira, and E. Cohler, *IEEE Trans. Magn.* **MAG-6**, 475 (1970).

34. K. D. Broadbent, *IRE Trans. Electron. Comput.* **EC-9**, 321 (1960).
35. A. C. Tickle, *J. Appl. Phys.* **35**, 768 (1964).
36. R. J. Spain, *IEEE Trans. Magn.* **MAG-2**, 347 (1966).
37. R. J. Spain, *IEEE Trans. Magn.* **MAG-3**, 334 (1967).
38. R. J. Spain and M. Marino, *IEEE Trans. Magn.* **MAG-6**, 451 (1970).
39. R. J. Spain and H. I. Jauvits, *Proceedings of the Intermagnetic Conference*, Paper 6-4, IEEE Press, New York (1974).
40. C. Batterel and M. Hanaut, *I.E.R.E. Conf. Proc.*, No. 35, 63 (1976).
41. D. O. Smith, *IRE Trans. Electron. Comput.* **EC-10**, 708 (1961); J. M. Ballantyne, *J. Appl. Phys.* **33**, 1067S (1962).
42. W. F. Druyvesteyn, A. W. M. vd Enden, F. A. Kuypers, E. de Niet, and A. G. H. Verhulst, in *Solid State Devices*, Proc. of the 4th European Solid State Device Research Conf. (ESSDERC), held at the University of Nottingham, September 16–19, 1973, p. 37, The Institute of Physics, London (1975).
43. A. H. Bobeck, *Bell Syst. Tech. J.* **46**, 1901 (1967).
44. A. A. Thiele, *Bell Syst. Tech. J.* **50**, 725 (1971).
45. U. F. Gianola, D. H. Smith, A. A. Thiele, and L. G. Van Uitert, *IEEE Trans. Magn.* **MAG-5**, 558 (1969).
46. H. Chang, *Magnetic Bubble Technology: Integrated Circuit Magnetics for Digital Storage and Processing*, IEEE Press, New York (1975).
47. P. Chaudhari, J. J. Cuomo, and R. J. Gambino, *IBM J. Res. Dev.* **17**, 66 (1973).
48. P. Chaudhari and S. R. Herd, *IBM J. Res. Dev.* **20**, 102 (1976).
49. J. W. Schneider, *IBM J. Res. Dev.* **19**, 587 (1975).
50. R. Hasegawa and R. C. Taylor, *J. Appl. Phys.* **46**, 3606 (1975).
51. A. H. Bobeck and H. E. D. Scovil, *Sci. Am.* **224**, 78 (June 1971); P. Chaudhari, J. J. Cuomo, R. J. Gambino, and E. A. Giess, on *Physics of Thin Films*, Vol. 9, G. Hass and M. H. Francombe, eds., Academic Press, New York (1977), p. 263.
52. E. P. Valstyn and L. F. Shew, *IEEE Trans. Magn.* **MAG-9**, 317 (1973).
53. L. T. Romanikiw, I. M. Croll, and M. Hatzakis, *IEEE Trans. Magn.* **MAG-6**, 597 (1970).
54. A. D. Kaske, P. E. Oberg, M. C. Paul, and G. F. Sauter, *IEEE Trans. Magn.* **MAG-7**, 675 (1971).
55. J. P. Lazzari, 19th AIP Conf. Magnetism and Magnetic Materials, Boston (1973).
56. J. P. Lazzari and I. Melnick, *IEEE Trans. Magn.* **MAG-6**, 601 (1970).
57. M. H. Hanazono, K. Kawakami, S. Narishige, O. Asai, E. Kaneko, K. Okuda, K. Ono, H. Tsuchiya, and W. Hayakama, *IEEE Trans. Magn.* **MAG-15**, 1616 (1979).
58. R. P. Hunt, *IEEE Trans. Magn.* **MAG-7**, 150 (1971).
59. Nobuo Kotera, Jungi Shigeta, Koziro Narita, Tetsu Oi, Kenji Kayashim and Kikuji Sato, *IEEE Trans. Magn.* **MAG-15**, 1946 (1979).
60. Lo-D Catalog on Cassette Tape Deck D-7500 Using Hall Effect Magnetic Heads, Hitatchi Ltd. (1978).
61. R. J. Spain, H. W. Fuller, and R. J. Webber, *IEEE Trans. Magn.* **MAG-2**, 288 (1966).
62. S. Sugatani, S. Konishi, and Y. Sakurai, *IEEE Trans. Magn.* **MAG-5**, 464 (1969).

References for Chapter 6

1. H. K. Onnes, *Leiden Commun.* **119b, 122b** (1911).
2. G. Rickayzen, *Theory of Superconductivity*, Wiley-Interscience, New York (1965).
3. A. C. Rose-Innes and E. H. Rhoderick, *Introduction to Superconductivity*, Pergamon Press, Oxford (1969).
4. M. Tinkham, *Introduction to Superconductivity*, McGraw-Hill, New York (1975).

5. J. R. Schrieffer, *Theory of Superconductivity*, W. A. Benjamin, New York (1964).
6. K. L. Chopra, *Thin Film Phenomena*, McGraw-Hill, New York (1969).
7. W. Meissner and R. Ochenfeld, *Naturwissenschaften* **21**, 787 (1933).
8. F. B. Silsbee, *J. Wash. Acad. Sci.* **6**, 597 (1916).
9. F. London, *Superfluids*, Vol. 1, Wiley, New York (1950).
10. F. London and H. London, *Proc. R. Soc. London, Ser. A* **149**, 71 (1935).
11. V. L. Ginzburg and L. D. Landau, *Zh. Eksp. Teor. Fiz.* **20**, 1064 (1950).
12. A. B. Pippard, *Proc. R. Soc. London, Ser. A* **216**, 547 (1953).
13. A. A. Abrikosov, *Zh. Eksp. Teor. Fiz.* **32**, 1442 (1957); *Sov. Phys. JETP (Engl. Transl.)* **5**, 1174 (1957).
14. H. Fröhlich, *Phys. Rev.* **79**, 845 (1950); *Proc. R. Soc. London, Ser. A* **228**, 296 (1954).
15. J. Bardeen, L. N. Cooper, and J. R. Schrieffer, *Phys. Rev.* **104**, 1175 (1957).
16. R. H. Blumberg and D. P. Seraphin, *J. Appl. Phys.* **33**, 163 (1962).
17. W. Buckel and R. Hilsch, *Z. Phys.* **138**, 109 (1954).
18. B. G. Lazarev, A. I. Sudovtsev, and A. P. Smirnov, *Sov. Phys. JETP (Engl. Transl.)* **6**, 816 (1958).
19. M. Tinkham, *Phys. Rev.* **129**, 2413 (1963).
20. J. W. Bremer, *Superconducting Devices*, McGraw-Hill, New York (1962).
21. V. L. Newhouse, J. W. Bremer, and H. H. Edwards, *Proc. IRE* **48**, 1395 (1960).
22. A. E. Brenneman, J. J. McNichol, and D. P. Seraphin, *Proc. IEEE*, **51**, 1009 (1963).
23. V. L. Newhouse and H. H. Edwards, *Proc. IEEE* **52**, 1191 (1964).
24. V. L. Newhouse, in *Treatise on Superconductivity*, R. Parks, ed., Marcel Dekker, New York (1969).
25. W. S. Goree and V. W. Hesterman, in *Applied Superconductivity*, Vol. 1, V. L. Newhouse, ed., Academic Press, New York (1975), p. 168.
26. V. L. Newhouse, J. L. Mundy, R. E. Joynson, and W. H. Meiklejohn, *Rev. Sci. Instrum.* **38**, 798 (1967).
27. M. K. Heynes, *Solid State Electron.* **1**, 399 (1960).
28. E. C. Crittenden, Jr., in *Proceedings of the 5th International Conference on Low-Temperature Physics and Chemistry*, J. R. Dillinger, ed., University of Wisconsin Press, Madison (1958), p. 232.
29. M. J. Buckingham, in *Proceedings of the 5th International Conference on Low-Temperature Physics and Chemistry*, J. R. Dillinger, ed., University of Wisconsin Press, Madison (1958), p. 229.
30. J. W. Crowe, *IBM J. Res. Dev.* **1**, 295 (1957).
31. E. C. Crittenden, Jr., J. N. Copper, and F. W. Schmidlin, *Proc. IRE* **48**, 1233 (1960).
32. C. R. Vail, M. S. P. Lucas, H. A. Owen, and W. C. Stewart, *Solid State Electron.* **1**, 279 (1960).
33. A. R. Sass, W. C. Stewart, and L. S. Cosentino, in *Applied Superconductivity*, Vol. 1, V. L. Newhouse, ed., Academic Press, New York (1975), p. 232.
34. R. A. Gange, *Electronics*, 111 (April 17, 1967).
35. R. F. Broom and O. Simpson, *Br. J. Appl. Phys.* **11**, 78 (1960).
36. E. H. Rhoderick, *Proc. R. Soc. London, Ser. A* **267**, 231 (1962).
37. T. A. Buchold, *Sci. Am.* **202**, 74 (1960); *Cryogenics* **1**, 203 (1961).
38. I. Giaever, *Phys. Rev. Lett.* **5**, 147 (1960).
39. I. Giaever, *Phys. Rev. Lett.* **5**, 464 (1960).
40. J. Nicol, S. Shapiro, and P. H. Smith, *Phys. Rev. Lett.* **5**, 461 (1960).
41. J. L. Miles, P. H. Smith, and W. Schönbein, *Proc. IEEE* **51**, 937 (1963).
42. L. Solymar, *Superconductive Tunnelling and Applications*, Chapman and Hall, London (1972).
43. B. D. Josephson, *Phys. Lett.* **1**, 251 (1962); *Rev. Mod. Phys.* **36**, 216 (1964); *Adv. Phys.* **14**, 419 (1965).

44. P. W. Anderson and A. H. Dayem, *Phys. Rev. Lett.* **13**, 195 (1964).
45. H. A. Notarys, R. H. Wang, and J. E. Mercereau, *Proc. IEEE* **61**, 79 (1973).
46. W. C. Stewart. *Appl. Phys. Lett.* **12**, 277 (1968).
47. D. E. McCumber, *J. Appl. Phys.* **39**, 3113 (1968).
48. T. F. Finnegan, A. Denenstein, and D. N. Langenberg, *Phys. Rev. B* **4**, 1487 (1971).
49. B. B. Schwartz and S. Foner, eds., *Superconductor Applications: SQUIDs and Machines*, Plenum Press, New York (1977).
50. J. E. Mercereau, *Rev. Phys. Appl.* **5**, 13 (1970).
51. H. A. Notarys and J. E. Mercereau, *J. Appl. Phys.* **44**, 1821 (1973); *J. Vac. Sci. Technol.* **10**, 646 (1973); R. K. Kirschman, H. A. Notarys, and J. E. Mercereau, *IEEE Trans. Magn.* **MAG-11**, 778 (1975).
52. T. Fujita, S. Kosaka, T. Ohtsuka, and Y. Onodera, *IEEE Trans. Magn.* **MAG-11**, 739 (1975).
53. J. Clarke, W. M. Goubau, and M. B. Ketchen, *IEEE Trans. Magn.* **MAG-11**, 724 (1975).
54. J. Clarke, W. M. Goubau, and M. B. Ketchen, *Appl. Phys. Lett.* **27**, 155 (1975).
55. M. B. Ketchen, W. M. Goubau, J. Clarke, and G. B. Donaldson, *IEEE Trans. Magn.* **MAG-13**, 372 (1977).
56. D. Cohen, *Phys. Today* **28**, 34 (August 1975).
57. S. J. Williamson, L. Kaufman, and D. Brenner, in *Superconductor Applications: SQUIDs and Machines*, B. B. Schwartz and S. Foner, eds., Plenum Press, New York (1977).
58. D. Cohen, *IEEE Trans. Magn.* **MAG-11**, 694 (1975).
59. D. Cohen, *Science* **180**, 745 (1973)
60. E. C. Hirschkoff, O. G. Symko, and J. C. Wheatley, *J. Low Temp. Phys.* **5**, 155 (1971).
1. E. C. Hirschkoff, O. G. Symko, L. L. Vant-Hall, and J. C. Wheatley, *J. Low Temp. Phys.* **2**, 653 (1970).
62. R. A. Webb, R. P. Giffard, and J. C. Wheatley, *J. Low Temp. Phys.* **13**, 383 (1973),
63. S. P. Boughn, M. S. McAshan, R. C. Taber, W. M. Fairbank, and R. P. Giffard, *Proceedings of the 14th International Conference on Low-Temperature Physics*, North Holland-American Elsevier (1975).
64. W. M. Wynn, C. P. Frahm, P. J. Carroll, R. H. Clark, J. Wellhoner, and M. J. Wynn, *IEEE Trans. Magn.* **MAG-11**, 701 (1975).
65. C. M. Swift, A Magnetotelluric Investigation of an Electrical Conductivity Anomaly in the S. W. United States, Ph.D. Thesis (unpublished), Geophysics Department, Massachusetts Institute of Technology, Cambridge, Massachusetts (1967).
66. M. J. S. Johnston and F. D. Stacey, *Nature* **224**, 1289 (1969).
67. R. A. Kamper, *IEEE Trans. Magn.* **MAG-11**, 141 (1975).
68. A. H. Silver, *IEEE Trans. Magn.* **MAG-15**, 268 (1979).
69. J. Clarke, in *Superconductor Applications: SQUIDs and Machines*, B. B. Schwartz and S. Foner, eds., Plenum Press, New York (1977).
70. J. Clarke, *Proc. IEEE* **61**, 8 (1973).
71. J. E. Zimmerman, *Cryogenics* **12**, 19 (1972).
72. P. L. Richards, T. M. Shen, R. E. Harris, and F. L. Lloyd, *Appl. Phys. Lett.* **36**, 480 (1980).
73. M. Nisenoff, *Rev. Phys. Appl.* **9**, 65 (1974).
74. T. M. Shen, P. L. Richards, R. E. Harris, and F. L. Lloyd, *Appl. Phys. Lett.* **36**, 777 (1980).
75. C. D. Tesche and J. Clarke, *J. Low Temp. Phys.* **29**, 301 (1977).
76. J. J. Matisoo, *Proc. IEEE* **55**, 172 (1967).
77. P. Gueret, *Appl. Phys. Lett.* **25**, 426 (1974).
78. J. J. Matisoo, *IEEE Trans. Magn.* **MAG-5**, 848 (1969).
79. W. Anacker, *Proceedings of the Fall Joint Computer Conference 41*, AFIPS Press (1972), p. 1269.
80. C. L. Huang and T. Van Duzer, *Appl. Phys. Lett.* **25**, 753 (1974).
81. T. R. Gheewala, *IEEE J. Solid-State Circuits* **SC-14**, 787 (1979).

82. W. Anacker, *IBM J. Res. Dev.* **24**, 107 (1980).
83. D. J. Herrell, *IEEE J. Solid-State Circuits* **SC-10**, 360 (1975).
84. Y. L. Yao and D. J. Herell, Int. Electron Devices Meeting Tech. Digest (1974), p. 145.
85. H. H. Zappe, *Appl. Phys. Lett.* **25**, 424 (1974).
86. H. H. Zappe, *Appl. Phys. Lett.* **27**, 432 (1975).
87. H. H. Zappe, *IEEE J. Solid-State Circuits* **SC-10**, 12 (1975).
88. R. F. Broom, W. Jutzi, and Th. O. Mohr, *IEEE Trans. Magn.* **MAG-11**, 755 (1975).
89. W. H. Henkels and H. H. Zappe, *IEEE J. Solid-State Circuits* **SC-13**, 591 (1978).
90. T. R. Gheewala, *IEEE Trans. Electron Devices* **ED-27**, 1857 (1980).
91. D. G. McDonald, *Phys. Today* **34**, 36 (February 1981); J. Matisoo, *Sci. Am.* **242**, 50 (May 1980).
92. J. Clarke, *Phys. Today* **24**, 30 (August 1971).
93. W. Anacker, in *Solid State Devices*, Proceedings of the 6th European Solid State Device Research Conference (ESSDERC), Munich, Sept. 13–16, 1976, p. 39, The Institute of Physics, Bristol (1977).
94. R. A. Kamper, *Proceedings of the Symposium on the Physics of Superconducting Devices*, Charlottesville, Paper M1, Office of Naval Research (1967).
95. R. A. Kamper and J. E. Zimmerman, *J. Appl. Phys.* **42**, 132 (1971).
96. A. H. Silver and J. E. Zimmerman, in *Applied Superconductivity*, Vol. 1, V. L. Newhouse, ed., Academic Press, New York (1975), p. 2; P. M. Richards and T. M. Shen, *IEEE Trans. Electron Devices* **ED-27**, 1909 (1980).
97. A. H. Dayem, B. I. Miller, and J. J. Wiegand, *Phys. Rev. B* **3**, 2949 (1971).
98. J. A. Hoyle, R. R. Humphris, and J. W. Boring, *IEEE Trans. Magn.* **MAG-11**, 690 (1975).
99. M. Cavallini, G. Gallinaro, and G. Scoles, *Z. Naturforsch, Teil A* **24**, 1850 (1969).
100. G. H. Wood and B. L. White, *Appl. Phys. Lett.* **15**, 237 (1969).

References for Chapter 7

1. K. L. Chopra and D. K. Pandya, *Thin Solid Films* **50**, 81 (1978).
2. E. H. Putley, in *Topics in Applied Physics*, Vol. 19: *Optical and Infrared Detectors*, R. J. Keyes, ed., Springer-Verlag, Berlin (1977), p. 82.
3. K. L. Chopra, *Thin Film Phenomena*, McGraw-Hill, New York (1969), p. 598.
4. K. L. Chopra and S. K. Bahl, *Phys. Rev. Sect. B* **1**, 2545 (1970).
5. K. L. Chopra and S. K. Bahl, *J. Appl. Phys.* **40**, 4171 (1969).
6. A. David Pearson, in *Modern Aspects of the Vitreous State*, Vol. 3, J. D. MacKensie, ed., Butterworths, Washington (1964), p. 29.
7. S. G. Bishop and W. J. Moore, *Appl. Opt.* **12**, 80 (1973).
8. T. D. Moustakas and G. A. N. Connell, *J. Appl. Phys.* **47**, 1322 (1976).
9. K. Yoshihara, *Jpn. J. Appl. Phys.* **14**, Suppl. 14-1 (1975).
10. J. J. Brissot, F. Desvignes, and R. Martres, *IEEE Trans. Electron. Devices* **20**, 613 (1978).
11. J. J. Brissot and R. Martres, *Ann. Chin. (Paris)* **10**, 185 (1975).
12. G. A. Zaitsev, V. G. Stashkov, and I. A. Krebtov, *Cryogenics* **16**, 440 (1976).
13. G. Gallinaro and R. Varona, *Cryogenics* **15**, 292 (1975).
14. M. Chester, *Phys. Rev.* **145**, 76 (1966).
15. R. J. Von Gutfeld, A. H. Nethercot, Jr., and J. A. Armstrong, *Phys. Rev.* **142**, 247 (1966).
16. M. Melloni, *Ann. Phys.* **28**, 371 (1833).
17. K. L. Chopra and P. Nath, *Phys. Status Solidi A* **33**, 333 (1976).
18. S. K. Barthwal and K. L. Chopra, *Phys. Status Solidi A* **36**, 533 (1976).

19. K. L. Chopra, A. P. Thakoor, S. K. Barthwal, and P. Nath, *Phys. Status Solidi A* **40**, 247 (1977).

20. K. L. Chopra, S. K. Bahl, and M. R. Randlett, *J. Appl. Phys.* **39**, 1525 (1968).

21. R. K. Willardson and A. C. Beer, eds., *Semiconductors and Semimetals*, Vol. 5: Infrared Detectors, Academic Press, New York (1970); Vol. 12: Infrared Detectors (II), Academic Press, New York (1977).

22. D. Chen, G. N. Otto, and F. M. Schmit, *IEEE Trans. Magn.* **MAG-9**, 66 (1973).

23. W. R. Harding, C. Hilsum, and D. C. Northrop, *Nature* **181**, 691 (1958).

24. K. Oe, Y. Toyoshima, and N. Nagai, *J. Non-Cryst. Solids* **20**, 405 (1976).

25. K. L. Chopra, P. K. Bhat, B. Singh, and S. Rajagopalan, *Solid State Commun.* **29**, 167 (1979); *J. Non-Cryst. Solids* **35**, 1053 (1980).

26. T. Igo and Y. Toyoshima, *J. Non-Cryst. Solids* **11**, 304 (1972).

27. A. Yoshikawa, O. Ochi, H. Nagai, and Y. Mizushima, *Appl. Phys. Lett.* **29**, 677 (1976).

28. A. Hadni, *J. Phys.* **24**, 694 (1963).

29. A. Hadni, Y. Henninger, R. Thomas, P. Vergnat, and B. Wyncke, *J. Phys.* **26**, 345 (1965).

30. B. O. Seraphin, in *Topics in Applied Physics*, Vol. 31: *Solar Energy Conversion*, B. O. Seraphin, ed., Springer-Verlag, Berlin (1979), p. 5.

31. R. N. Schmidt, *J. Spacecr. Rockets* **2**, 101 (1965).

32. B. O. Seraphin and A. B. Meinel, in *Optical Properties of Solids: New Developments*, B. O. Seraphin, ed., North-Holland, Amsterdam (1976), p. 927.

33. L. E. Flordal and R. Kivaisi, *Vacuum* **27**, 397 (1977).

34. T. J. McMahon and S. W. Jasperson, *Appl. Opt.* **13**, 2750 (1974).

35. D. K. Pandya and K. L. Chopra, in *Vacuum–Surfaces–Thin Films*, K. L. Chopra and T. C. Goel, eds., Vanity Books, Delhi (1971).

36. D. C. Booth, D. D. Allerd, and B. O. Seraphin, *Solar Energ. Mater.* **2**, 107 (1979).

37. R. B. Goldner and H. M. Haskal, *Appl. Opt.* **14**, 2328 (1975).

38. S. N. Kumar, L. K. Malhotra, and K. L. Chopra, *Solar Energ. Mater.* **3**, 519 (1980).

39. P. K. Gogna, D. K. Pandya, and K. L. Chopra, *Proceedings of the International Solar Energy Conference*, New Delhi (1978), p. 842.

40. P. K. Gogna and K. L. Chopra, *Thin Solid Films* **63**, 183 (1979); P. K. Gogna, Ph.D. Thesis, Indian Institute of Technology, Delhi (1980).

41. R. N. Schmidt, K. C. Park, and J. E. Janssen, Tech. Report, Wright–Patterson Air Force Base, ML-TDR-64-250 (1964).

42. G. B. Reddy, V. Dutta, D. K. Pandya, and K. L. Chopra, *Solar Energ. Mater.* **5**, 187 (1981).

43. S. N. Kumar, L. K. Malhotra, and K. L. Chopra, communicated to *Solar Energ. Mater.*

44. D. A. Williams, T. Lappin, and A. J. Duffie, *Trans. ASME Ser. A, J. Eng. Power* (July 1963), p. 213; B. K. Gupta, R. Thangraj, and O. P. Agnihotri, *Solar Energ. Mater.* **1**, 48 (1979).

45. B. O. Seraphin, in *Proceedings of the Symposium on the Material Science Aspects of Thin Films Systems for Solar Energy Conversion*, B. O. Seraphin, ed., Tucson, Arizona, July 1974.

46. G. D. Pettit, J. J. Cuomo, T. H. DiStefano, and J. M. Woodale, *IBM J. Res. Dev.* **22**, 372 (1978).

47. C. M. Horwitz, *Opt. Commun.* **11**, 210 (1974).

48. D. Pramanik, A. J. Sievers, and R. H. Silsbee, *Solar Energ. Mater.* **2**, 81 (1979).

References for Chapter 8

1. P. F. Kane and G. B. Larrabee, *Characterization of Solid Surfaces*, Plenum Press, New York (1974).

2. A. W. Czanderna, ed., *Methods of Surface Phenomena*, Vol. 1, Elsevier, Amsterdam (1975).

3. L. H. Lee, ed., *Characterization of Metal and Polymer Surfaces*, Vols. 1 and 2, Academic Press, New York (1977).

4. F. R. Eirich, in *Interface Conversion for Polymer Coatings*, P. Weiss and G. Dale Cheever, eds., American Elsevier, New York (1968), p. 351.

5. C. A. Krier, in *Vapor Deposition*, C. F. Fowell, J. H. Oxley, and J. M. Blocher, Jr., eds., Wiley, New York (1966), p. 512.

6. R. M. Burns and W. W. Bradley, *Protective Coatings for Metals*, 2nd edn., Reinhold, New York (1955).

7. O. Kubaschewski and B. E. Hopkins, *Oxidation of Metals and Alloys*, 2nd edn., Academic Press, New York (1962).

8. G. T. Bakhvalov and A. V. Turkovskaya, *Corrosion and Protection of Metals*, translated and edited by G. Isserli, Pergamon Press, Oxford (1965).

9. U. R. Evans, *An Introduction to Metallic Corrosion*, Edward Arnold, London (1960).

10. S. Remalingam, Y. Shimazaki, and W. O. Winer, *Thin Solid Films* **80**, 297 (1981).

11. R. F. Christy, *Thin Solid Films* **80**, 289 (1981).

12. E. Lenz, D. Pneuli, and L. Rozeanu, *Wear* **53**, 337 (1979).

13. N. Ohmae, T. Tsukizoe, and T. Nakal, *J. Lubr. Technol.* **100**, 129 (1978); V. D. Vankar and K. L. Chopra, in *Vacuum–Surfaces–Thin Films*, K. L. Chopra and T. C. Goel, eds., Vanity Books, Delhi (1981).

14. W. Schintlmeister and O. Pacher, *J. Vac. Sci. Technol.* **12**, 743 (1975).

15. H. E. Hintermann, H. Boring, and W. Hanni, *Wear* **48**, 225 (1975).

16. T. A. Wolfla and R. C. Tucker, Jr., *Thin Solid Films* **53**, 353 (1978).

17. N. V. Novikov and P. S. Kisly, *Thin Solid Films* **64**, 205 (1979).

18. W. E. Jamison, *Thin Solid Films* **73**, 227 (1980).

19. R. L. Johnson, *Thin Solid Films* **73**, 253 (1980).

20. R. G. Duckworth, *Thin Solid Films*, **73**, 275 (1980).

21. R. F. Bunshah and D. M. Mattox, *Phys. Today* **33**, 50 (May 1980).

22. E. Rabinowicz, *Friction and Wear of Materials*, John Wiley, New York (1965).

23. Koreo Kinosita, *Thin Solid Films* **12**, 17 (1972).

24. R. W. Hoffman, in *Physics of Thin Films*, Vol. 3, G. Hass and R. E. Thun, eds., Academic Press, New York (1966), p. 211.

25. R. F. Bunshah, *J. Vac. Sci. Technol.* **11**, 633 (1974); *Vacuum* **27**, 353 (1977); B. E. Jacobson, R. F. Bunshah, and R. Nimmagadda, *Thin Solid Films* **54**, 107 (1978).

26. K. Nakamura, K. Inagawa, K. Tsuruoka, and S. Komiya, *Thin Solid Films* **40**, 155 (1977).

27. H. Ortner and K. A. Gebler, U.S. Dept. Commerce Office Tech. Serv. AD Rep. 266711 (October 1961); AD Rep. 285514 (July 1962).

28. M. Kodama, A. H. Shabaik, and R. F. Bunshah, *Thin Solid Films* **54**, 353 (1978).

29. F. P. Bowden and D. Tabor, *The Friction and Lubrication of Solids*, Clarendon Press, Vol. 1 (1950); Vol. 2 (1964).

30. T. Spalvis, *Thin Solid Films* **53**, 285 (1978).

31. H. E. Sliney, *Thin Solid Films* **64**, 211 (1979).

32. F. J. Horn and H. B. Hebble, Jr., in *Vapor Deposition*, C. F. Powell, J. H. Oxley, and J. M. Blocher, Jr., eds., Wiley, New York (1966), p. 579.

33. F. A. Lowenheim, *Electroplating*, McGraw-Hill, New York (1978), p. 188.

34. K. L. Mittal and D. F. O'Kane, *J. Adhes.* **8**, 93 (1976).

35. I. Haller, *J. Am. Chem. Soc.* **100**, 8050 (1978).

36. F. M. Fowkes, *J. Adhes.* **4**, 155 (1972).

37. F. M. Fowkes and S. Maruchi, *Coat. Plast. Prepr.* **37**, 605 (1977).

38. E. P. Plueddemann, in *Interface Phenomena in Polymer Matrix Composite Materials*, E. P. Plueddemann, ed., Academic Press, New York (1974), p. 173.

39. B. C. Arkles, *Chem. Technol.* 766 (December 1977).
40. M. R. Rosen, *J. Coat. Technol.* **50**, 72 (September 1978).
41. G. C. Bond, *Catalysis by Metals*, Academic Press, London (1962), p. 29.
42. O. Beeck, A. E. Smith, and A. Wheeler, *Proc. R. Soc. London, Ser. A* **177**, 62 (1941).
43. G. C. Bond, *Trans. Faraday Soc.* **52**, 1235 (1956).
44. J. H. Oxley, in *Vapor Deposition*, C. F. Powell, J. H. Oxley, and J. M. Blocher, Jr., eds., Wiley, New York (1966), p. 484.
45. F. G. Foote and J. F. Schumer, USAEC Rep. ANL-4364 (September 30, 1949; declassified May 10, 1957).
46. A. P. Backensto, Jr., N. F. Hopson, and F. V. Lenel, USAEC Rep. SO-3005 (February 5, 1952; declassified January 31, 1956).
47. *Proc. of Symp. on Ceramic Matrix Fuels Containing Coated Particles*, held at BMI, November 5–6, USAEC Rep. TID-7654 (April 1963).
48. C. W. Townley, N. E. Miller, R. L. Ritzman, and R. J. Burian, USAEC Rep. BMI-1613 (January 30, 1963); USAEC Rep. BMI-1628 (April 25, 1963); *Nucleonics* **22**, 45 (February 1964).
49. A. K. Smalley, W. C. Riley, and W. H. Duckworth, USAEC Rep. BMI-1321 (February 18, 1959); A. K. Smalley, M. C. Brockway, and W. H. Duckworth, USAEC Rep. BMI-1579 (May 22, 1962); *Am. Ceram. Soc. Bull.* **42**, 494 (1963).
50. J. M. Blocher, Jr., in *Vapor Deposition*, C. F. Powell, J. H. Oxley, and J. M. Blocher, Jr., eds., Wiley, New York (1966), p. 650.
51. L. Meyer and R. Gomer, *J. Chem. Phys.* **28**, 617 (1958).
52. L. Mond, U.S. Patent No. 455,230 (June 30, 1891).
53. J. C. Ladd and D. L. Allie, *ASTM Spec. Tech. Publ.* **318**, 124 (1962).
54. J. A. M. Van Liempt, *Metallwirtschaft* **11**, 357 (1932).
55. H. C. Nathanson and J. Guldberg, in *Physics of Thin Films*, Vol. 8, G. Hass, M. H. Francombe, and R. W. Hoffman, eds., Academic Press, New York (1975), p. 251.
56. V. R. Brown and R. McCusker, *Proceedings of the 21st Annual Conference on Engineering in Medicine and Biology*, held at Houston, Texas (1968), Vol. 10, p. 13A5.
57. R. L. White and H. D. Mercer, in *Biomedical Electrode Technology*, H. A. Miller and D. C. Harrison, eds., Academic Press, New York (1974), p. 159.
58. K. D. Wise, J. B. Angell, and A. Starr, *IEEE Trans. Biomed. Eng.* **BME-17**, 238 (1970).
59. W. M. Portnoy, R. M. David, and L. A. Akers, in *Biomedical Electrode Technology*, H. A. Miller and D. C. Harrison, eds., Academic Press, New York (1974), p. 7.
60. L. F. Montes, J. L. Day, and L. Kennedy, *J. Invest. Dermatol.* **49**, 100 (1967); C. D. Wheelwright, *Techn. Note D-1082*, NASA MSC, Houston (1969).
61. P. W. Barth and J. B. Angell, *IEEE Trans. Electron Devices* **ED-29**, 144 (1982).
62. S. A. S. Donoghue, G. A. May, N. E. Cotter, R. L. White, and F. B. Simmons, *IEEE Trans. Electron Devices* **ED-29**, 136 (1982).
63. M. R. Havens, M. E. Biolsi, and K. G. Mayhan, *J. Vac. Sci. Technol.* **13**, 575 (1976).
64. P. J. Dynes and D. H. Kaelble, in *Plasma Chemistry of Polymers*, M. Shen, ed., Marcel Dekker, New York (1976), p. 167.
65. H. K. Yasuda, in *Contemporary Topics in Polymer Science*, Vol. 3, M. Shen, ed., Plenum Press, New York (1979), p. 103.
66. A. W. Hahn, M. F. Nicholos, A. K. Sharma, and E. W. Hellmuth, in *Biomedical and Dental Applications of Polymers*, C. G. Gebelein and F. K. Koblitz, eds., Plenum Press, New York (1981), p. 85.
67. R. K. Sadhir, W. J. James, H. K. Yasuda, A. K. Sharma, M. F. Nicholos, and A. W. Hahn, *Biomaterials* **2**, 239 (1981).
68. A. W. Hahn, K. G. Mayhan, J. R. Easley, and C. W. Sanders, *Natl. Bur. Stand. (U.S.) Spec. Publ.* 415, Biomaterials (May, 1975), p. 13.

69. A. S. Chawla, *Artif. Org.* **3**, 92 (1979).
70. K. G. Mayhan, A. W. Hahn, and R. B. Barr, *Proceedings of the 25th Annual Conference on Engineering in Medicine and Biology*, Miami Beach, Florida (1972), p. 85.
71. H. Yasuda, M. O. Bumgarner, H. C. Marsh, B. S. Yamanashi, M. P. Devito, M. L. Wolbarsht, J. W. Reed, M. Bessler, M. B. Landers, III, D. M. Hercules, and J. Carver, *J. Biomed. Mater. Res.* **9**, 629 (1975).
72. A. W. Hahn, H. K. Yasuda, W. J. James, M. F. Nicholos, R. K. Sadhir, A. K. Sharma, O. A. Pringle, D. H. York, and E. J. Carlson, *Biomedical Sciences Instrumentation*, Vol. 17, Proc. 18th Ann. Rocky Mountain Bioengineering Symp., Laramie, Wyoming, April 20–21, 1981, p. 109, Instrument Society of America, Pittsburgh (1981).

Index